Industrial IoT for Architects and Engineers

Architecting secure, robust, and scalable industrial IoT solutions with AWS

Joey Bernal

Bharath Sridhar

BIRMINGHAM—MUMBAI

Industrial IoT for Architects and Engineers

Group Product Manager: Rahul Nair
Publishing Product Manager: Niranjan Naikwadi
Senior Editor: Tanya D'cruz
Technical Editor: Rajat Sharma
Copy Editor: Safis Editing
Project Coordinator: Deeksha Thakkar
Proofreader: Safis Editing
Indexer: Hemangini Bari
Production Designer: Shyam Sundar Korumilli
Senior Marketing Coordinator: Nimisha Dua

First published: January 2023

Production reference: 1211222

Published by Packt Publishing Ltd.
Livery Place
35 Livery Street
Birmingham
B3 2PB, UK.

ISBN 978-1-80324-089-3

www.packt.com

Contributors

About the authors

Anthony (Joey) Bernal is a creative technical leader focused on the **Internet of Things (IoT)** and cloud architecture. He has led the development of two major commercial IoT platforms from conception to general availability. He built and ran an IoT start-up, recognized by both Fast Company and Gartner, with customers in manufacturing, oil and gas, and agriculture. Joey is a hands-on architect with solid experience in development, infrastructure, IoT hardware, and cloud and edge platforms. He is an experienced writer and presenter with leadership skills, flexibility, creativity, and technical know-how, which have led to the delivery of many successful products and projects and a sense of humor to enjoy still doing it.

Bharath Sridhar is a technology evangelist and solution architect with over 12 years of experience in digital transformation through IoT. With a constantly curious and exploratory mindset, he works as an enabler of industry 4.0 implementations for Fortune 500 companies. He loves to operate at the intersection of desirability, viability, and feasibility, working to create utilitarian solutions that people love and businesses get delighted and technologists get excited about. He is passionate about knowledge sharing through storytelling. He believes that books are a gateway to curated journeys of personal discovery, experiences, and enlightened knowledge. In his free time, he dreams about the multiverse – its evolution, challenges, and solutions.

I would like to express my gratitude and thanks to all the teachers who have imparted knowledge during my existence, giving it purpose and direction.

About the reviewer

André Hoettgen has been working in industrial manufacturing for more than 10 years and specializes in cross-functional software and solution development. As an **Industrial Internet of Things (IIoT)** architect, he builds and manages IT/OT infrastructures, develops strategic solutions, and leads their integration. In the wild IIoT world, he organizes vast amounts of data and processes and harmonizes the most diverse systems. His extensive knowledge in numerous disciplines enables him to meet all stakeholders at their level and deliver cross-industry innovations.

Table of Contents

Part 1: An Introduction to Industrial IoT and Moving Toward Industry 4.0

1

Welcome to the IoT Revolution   3

2

Anatomy of an IoT Architecture   23

3

In-Situ Environmental Monitoring   51

4

Real-World Environmental Monitoring   73

Part 2: IoT Integration for Industrial Protocols and Systems

5

6

Preface

Anyone who picks up this book already knows something about the **Internet of Things (IoT)**. The term is everywhere, in the news and on social media, with **Industrial IoT (IIoT)** quickly becoming the focus for many organizations. But how do you get started? How do you design and build a system that will provide new levels of engagement with your environment? In addition, the level of noise in this field is astounding. Hundreds of vendors and solution providers are trying very hard to explain the right way (their way) of implementing IoT within your environment.

This book provides one such approach, with different options to build an IIoT ecosystem for your environment using Amazon Web Services. We will cut through the hype, getting to the heart of what is needed, why it is necessary, and how to approach your specific objectives. This book looks at designing a robust and scalable IoT platform, starting with the basics and adding capability along the way.

Our goal is to provide theory and hands-on examples for you to learn. We provide the information you need to make thoughtful choices and help your organization grow as you move further toward Industry 4.0. In addition, there are plenty of hands-on examples within these pages. These examples are simple enough that you can understand and implement the example code and scenarios but still complex enough to be considered jump-off points for your journey. This is not your father's "Hello, World!"

We hope you enjoy reading and learning from this book as much as we did writing it!

Who this book is for

This book is for architects, engineers, developers, and technical professionals interested in building an IoT ecosystem focused on industry solutions. This book will guide you in thinking about architecture and design that provides the integration and capability you need while focusing on good architecture as you go.

As readers, you may come from various industries looking to design and build a solution that can stand up to the rigors of an industrial or extreme environment. For this reason, the examples in this book will use readily available industrial-strength hardware and software so that you can recreate a robust solution that provides immediate value to your organization.

What this book covers

Chapter 1, Welcome to the IoT Revolution, focuses on Industry 4.0 and using data to drive efficiency and optimization. A data-driven mindset will be our *why*, and IoT will be the *how*. The goal of this chapter will be to understand some history and theory, think about your business cases, and start to define a solution in such a way that you can derive meaningful results while working to move your industry forward.

Chapter 2, Anatomy of an IoT Architecture, walks you through the decision-making process when designing a cloud application. We aim to help you think like an IT architect. Where does the process begin, and how do we accomplish our business goals from a technical perspective? We will evaluate options and trade-offs as we review the architectural layers within the overall design.

Chapter 3, In-Situ Environmental Monitoring, looks at environmental monitoring solutions more holistically. In this chapter, we will explore several common industries and use cases for environmental data collection. We will focus on the approach and the value we can get from our data, looking at how we can collect measurements from various circumstances outside traditional machine data capture.

Chapter 4, Real-World Environmental Monitoring, provides a look at environmental monitoring solutions, which have slightly different goals than what might be considered traditional industrial monitoring. For example, agriculture and ranching are diverse industries that depend significantly on environmental factors. The oil, gas, mining, and maritime sectors also have potential use cases.

Chapter 5, OT and Industrial Control Systems, considers existing control systems prevalent in the industry, commonly attributed as the OT layer. These are part of layers 0, 1, and 2 of the Purdue model and ISA-95 architecture. This chapter will help us understand real-time manufacturing execution systems and how machines are orchestrated to maximize production.

Chapter 6, Enabling Industrial IoT, logically forms the crucial integration layer for enabling IIoT applications. The objective of this chapter is to dive deep into the complexities and nuances to appreciate the need for such convergence between IT systems and OT systems. While you will be aware of the challenges of integration in a typical manufacturing scenario, you can also take away strategies and ideas to enable their convergence.

Chapter 7, PLC Data Acquisition and Analysis, is a hands-on chapter that will help you understand and appreciate an actual integration of a **Programmable Logic Controller** (**PLC**) with a data acquisition system. It starts with the architecture and design of a programmable logic controller and the evolution of hardware from the 1960s. We will also introduce you to programming a PLC with ladder logic and tools from various **Original Equipment Manufacturer** (**OEMs**). The critical facet of configuring the protocols, mapping the tags, and retrieving the data from PLCs shows you a fundamental integration point between OT and IT.

Chapter 8, Asset and Condition Monitoring, explores the ability to monitor and assess the performance of your equipment and processes across the factory floor. Unhealthy assets can contribute to various issues across the manufacturing process, from unexpected downtime to reduced productivity and output quality. Our goal is to contribute to efficiency by monitoring, maintaining, and improving each piece of equipment within a process.

Chapter 9, Taking It Up a Notch – Scalable, Robust, and Secure Architectures, looks at architectures fundamental to building a system. This chapter will address the need for an architectural framework to develop scalable, secure, and robust IIoT applications. This chapter will cover the broad spectrum of industry-wide architecture IIoT design considerations, with references from the **Industrial Internet Consortium** (**IIC**) and Reference Architecture Model Industrie 4.0.

Chapter 10, Intelligent Systems at the Edge, focuses on *edge technology* – compute power located within the factory. While we will still need to leverage the cloud and send data for processing and storage, our immediate activity will be more local so that we can be located near our data and the systems with which we interact.

Chapter 11, Remote Monitoring Challenges, are fundamental to autonomous operations. Remote monitoring plays a significant role in business continuity, predicting and mitigating failure scenarios. This chapter will focus on remote situations and decision-making about bandwidth concerns, power consumption, and the volume of data. We will examine different options for data transfer, such as 5G, satellite, and long-range wireless options.

Chapter 12, Advanced Analytics and Machine Learning, provides a comprehensive machine learning example. We explain how to derive an anomaly model based on sample data collected from the edge. Our primary goal is to illustrate the end-to-end, model-driven data engineering processes for our IoT efforts.

To get the most out of this book

We ask that you bring an open mind and have some patience. Not every step within the examples is documented. We expect you to have some technical skills in the cloud, AWS, and solution architecture. Additionally, as we point out in several chapters, if something appears not to be working, it is most likely due to AWS permissions.

Software/hardware covered in the book	Operating system requirements
Python: Knowledge of Python will be helpful. Code examples are in Python but are easy to follow if you are unfamiliar with the language. The code samples are easy to read and edit in your environment.	Much of the hands-on work is done within the AWS console. You will require an AWS account and some familiarity. Most examples work within the free tier; however, in the final chapter, which uses Amazon SageMaker, you should track your cost closely.
Visual Studio Code: This is used in some of the later chapters for all the Python code examples. This includes lambdas and edge components that we create. As an IDE, it is easy to use and free!	All the edge processing examples within the book use the Linux operating system, which may require knowing the basic commands when working on a Linux system.
Edge device: An edge device for installing and running Greengrass will be necessary. A Raspberry Pi will do the trick if you can find one. It should be on the same network as your Modbus simulator or device. This simulator can be a Windows edge device or PC if you are more comfortable with that OS.	

Download the example code files

You can download the example code files for this book from GitHub at `https://github.com/PacktPublishing/Industrial-IoT-for-Architects-and-Engineers`. If there's an update to the code, it will be updated in the GitHub repository.

We have code bundles from our rich catalog of books and videos available at `https://github.com/PacktPublishing/`. Check them out!

Download the color images

We also provide a PDF file that has color images of the screenshots and diagrams used in this book. You can download it here: `https://packt.link/wi9wN`

Conventions used

There are several text conventions used throughout this book.

`Code in text`: Indicates code words in text, database table names, folder names, filenames, file extensions, pathnames, dummy URLs, user input, and Twitter handles. Here is an example: "This `WirelessDeviceId` is created by AWS when you add the device."

A block of code is set as follows:

```
def lambda_handler(event, context):


    print("Received event: " + json.dumps(event, indent=2))
```

Any command-line input or output is written as follows:

```
$ cd /greengrass/v2/logs/
$ tail -f com.environmentsense.modbus.ModbusRequest.log
```

Bold: Indicates a new term, an important word, or words that you see on screen. For instance, words in menus or dialog boxes appear in **bold**. Here is an example: "Select **System info** from the **Administration** panel."

> Tips or important notes
> Appear like this.

Get in touch

Feedback from our readers is always welcome.

General feedback: If you have questions about any aspect of this book, email us at `customercare@packtpub.com` and mention the book title in the subject of your message.

Errata: Although we have taken every care to ensure the accuracy of our content, mistakes do happen. If you have found a mistake in this book, we would be grateful if you would report this to us. Please visit `www.packtpub.com/support/errata` and fill in the form.

Piracy: If you come across any illegal copies of our works in any form on the internet, we would be grateful if you would provide us with the location address or website name. Please contact us at `copyright@packt.com` with a link to the material.

If you are interested in becoming an author: If there is a topic that you have expertise in and you are interested in either writing or contributing to a book, please visit `authors.packtpub.com`.

Share Your Thoughts

Once you've read *Industrial IoT for Architects and Engineers*, we'd love to hear your thoughts! Scan the QR code below to go straight to the Amazon review page for this book and share your feedback.

`https://packt.link/r/180324089X`

Your review is important to us and the tech community and will help us make sure we're delivering excellent quality content.

Download a free PDF copy of this book

Thanks for purchasing this book!

Do you like to read on the go but are unable to carry your print books everywhere? Is your eBook purchase not compatible with the device of your choice?

Don't worry, now with every Packt book you get a DRM-free PDF version of that book at no cost.

Read anywhere, any place, on any device. Search, copy, and paste code from your favorite technical books directly into your application.

The perks don't stop there, you can get exclusive access to discounts, newsletters, and great free content in your inbox daily

Follow these simple steps to get the benefits:

1. Scan the QR code or visit the link below

https://packt.link/free-ebook/9781803240893

2. Submit your proof of purchase
3. That's it! We'll send your free PDF and other benefits to your email directly

Part 1: An Introduction to Industrial IoT and Moving Toward Industry 4.0

Need help with the basics of the **Industrial Internet of Things (IIoT)** and looking to learn more about its importance? The first two chapters will provide just that, introducing digital evolution and helping decision-makers realize business values. This section explains IIoT and enterprise architecture's critical core foundation components, relating them to the AWS cloud. This part of the book will also assist you in getting started with IoT and solution architecture principles.

The second half of this section takes you deeper into wireless technology approaches to IoT. This approach allows us to learn key concepts and build an end-to-end example of collecting data from the field. Architecture is an evolution, and we will explore many IoT concepts by starting here. Additionally, we will take a few baby steps in collecting, processing, storing, and displaying your IoT data.

This part of the book comprises the following chapters:

- *Chapter 1, Welcome to the IoT Revolution*
- *Chapter 2, Anatomy of an IoT Architecture*
- *Chapter 3, In-Situ Environmental Monitoring*
- *Chapter 4, Real-World Environmental Monitoring*

1
Welcome to the IoT Revolution

This book is designed for architects and industrial engineers looking for guidance in moving into IoT or the Industry 4.0 space, offering some ideas, approaches, goals, and advice to help make your way forward a little easier and more successful. For readers new to IT architecture or the IoT space, we aim to help answer many of those initial questions or at least guide you in asking the right questions. We want to set the stage in these initial chapters before we get too deep into the technical details. Anyone new to these topics should benefit from these initial chapters, especially non-technical stakeholders who want to understand the *why* and *how* of Industrial IoT. With this in mind, please consider that we are providing historical and architectural background, guidance, and some best practices, from an IT and system development approach. If things get too technical, you have been warned.

In this chapter, we want to set the stage for understanding Industry 4.0 and help you to understand where it is headed and why it is important. We are going to review why the current Industrial Revolution and Industry 4.0 are so important, know where we are in the current state of technology, and learn how you can build your vision and value statement for driving technologies such as IoT into your organization.

We are going to cover the following main topics:

- Industry 4.0 and the digitalization of industry
- How IoT can support Industry 4.0 at scale
- The convergence – IT, OT, and management working together
- Leveraging good architecture to drive progress

In future chapters, we will delve much deeper into the *how* of Industrial IoT and learn how to implement and use some of this exciting technology, but stay with us. We need to understand some of the history better and discover where it all started. We have also chosen **Amazon Web Services** (**AWS**) as the hyperscaler of choice to base our practical examples. AWS is a formidable player in this space and has a great product roadmap and vision associated with Industrial IoT. There will be more about this as we progress across chapters.

Technical requirements

There are no specific technical requirements for this chapter. Readers at every level should clearly understand it. Our focus is setting the groundwork for why Industrial IoT is poised as one of the next major turns of the technology crank and how you can move forward with adoption within your industry.

Industry 4.0 and the digitalization of industry

Many software architects are sometimes wary of the hype around new technology. Great ideas and visions are pivots that lead us into the future and guide us in taking advantage of new technology in both our business and personal lives. However, the road to the current state of technology is paved with great ideas that never made it out of the concept phase, and overly aggressive marketing and sales around new (good and bad) technologies have made everyone just a little more cautious.

Usually, at the early stages of some technologies, marketing and sales teams jump in and take over, looking for any opportunity to push an idea or build a prototype with any potential customer, attempting to work together with customers to build a vision of what the future could be. But then comes the hard work of architecture, design, prototyping, rollout, testing, production, and support. Sometimes, the state of the technology isn't quite ready, and reality intervenes. If you have been burned enough times, it gets harder to reach back in.

Fortunately for us, Industry 4.0 has made it well past the starting gate and into the reality of many organizations. Even though it has been making progress for most of the last decade, there is still a fair amount of work to be done before it can be considered mainstream technology in many organizations. The evolution and improvements in hardware, such as sensors and processors, software protocols, and integration tools, make retrieving real-time or near real-time data from almost any device or area more accessible and safer. The *why* of data capture and Industrial IoT is what we will be discussing in this chapter, while the *how* will be discussed in the rest of this book.

Industry 4.0, or the fourth Industrial Revolution, is commonly thought of as the automation and digitalization of industry and manufacturing systems. IoT and cloud technologies have become critical enablers of this effort and provide the ability to integrate and automate machinery to become more intelligent and adaptive. Ideally, this includes adopting artificial intelligence and machine learning to enable systems to self-monitor and diagnose or predict problems that may occur.

This description does provide a bit of futuristic vision, connotating a kind of *rise of the machines* approach, but it gives us a good starting point on which to base our discussion.

A very brief history lesson

History books and most university classes on this topic will agree that the world has undergone three previous industrial revolutions. For us, how we got to where we are is maybe not as important as where we are going, so we won't belabor the history here, but we'll provide some background to aid in your organizational discussions and help us pinpoint the reason for and the focus of this book.

The first Industrial Revolution

The first Industrial Revolution occurred in the late 1700s when mechanization based on water or steam power began. Traditional thought placed this beginning in the 1780s when the first mechanical loom was designed and built. While (relatively) easy to make, replicate, and ship, this allowed for the first major transition from production using hands to allowing machine-based tools to do the work.

Early industrial progress

There are, of course, precursors to the first Industrial Revolution. Recently, on a trip to the Netherlands, I was able to tour some windmills that advanced industry in the region as early as the 1600s, providing improvements to industries such as milling, weaving, and lumber production. Although windmill technology had been in service moving water in the region for centuries before this, this small evolution in leveraging the technology for other types of work allowed the Netherlands to advance into a new era, most notably in shipbuilding. Unfortunately, the technology could not be as easily exported since wind-driven machines were primarily a defining factor of the region. However, the inventiveness of the Dutch and the innovative use of gears, levers, and screws helped build the groundwork for future industry advances, evolving from, for example, farm animals for drawing water or agriculture.

The fact that much of the work was driven by steam was also important. The steam engine's efficiency had greatly improved by this time, and it was now lighter and more transportable. Coal, and the ability to mine coal in significant quantities, was essential for powering these steam engines. Adapting these same engines moving in one direction or performing one motion to a different degree of movement allowed for more flexibility and complexity in industrial use. The loom was prominent in this phase because the textile industry was labor-intensive, and it became one of the first industries to adopt and see the benefit of new technology.

The second Industrial Revolution

The second Industrial Revolution often referred to as the technological revolution, started in the late *1800s* and was a strong driver for the modern world we live in today. The expansion of almost everything we know and use in today's world started during this period. Beginning with the growth of railroads and telegraphs, industry expanded further, bringing gas, water, sewer, and electricity and increasing globalization toward the end of the colonial age.

The expansion of electricity and assembly/production lines happened within this period. History credits Henry Ford for inventing the assembly line in 1913, paving the way for advanced mass production. Ford is also credited for advancements with the combustion engine, steel, and new fuels and materials that drove this exciting period of change and once again transformed many industries.

The third Industrial Revolution

The timelines are a little intertwined because advancements were frequently made that lent toward each distinct phase of technological evolution. These revolutions can seem almost continuous if traced from end to end with enough detail and advancements. There have always been significant breakthroughs that highlight the end of the last and the beginning of the next phase of advancement. The third Industrial Revolution started in the late 1900s and is called the **Digital Revolution**. This registers as a shift from analog technology to digital technology. The invention of the internet and smaller computing technologies allowed us to enter the information age.

The invention of the **transistor** in 1947 is a critical starting point for this era. However, it was several decades before this technology was adapted enough to be helpful on a large scale, with the ability to design and create integrated circuits consisting of hundreds of transistors. Eventually, this allowed the creation of the single-chip microprocessor in 1971 by Intel, allowing for desktop computers to become readily available.

Moving forward and the fourth Industrial Revolution

Hopefully, this short history lesson about the previous three industrial revolutions has helped you understand where we started and assisted you in visualizing how the technology crank continuously turns. Before you know it, advancement has occurred. In addition, each revolution has added tremendous value, advancing civilization, increasing productivity and safety, and moving the entire world another step forward.

The fourth Industrial Revolution should have no less lofty goals, with even more of a potential impact on civilization as a whole. I admit this sounds a bit too rosy, but think about it in terms of the effects on humanity and the world we live in. Efficiency itself means less waste, less use of energy, and potentially less pollution and impact on the environment. That, in itself, should make an effort to move forward worthwhile, and that these improvements can help increase productivity, quality, and revenue is icing on the cake.

Keep this in mind as you delve through this book and determine how to apply some of the ideas to your industry. The immediate goal may be to save or make more money; however, inside, you should know that you are hopefully doing your small part to help save the world.

Achieving the vision of Industry 4.0 requires effort and time and cannot be completed all in one go. This is especially true for legacy or brownfield industrial operations that have sometimes been in service for decades. Additionally, some industries produce widely varied results based on external conditions, such as farming.

Earlier in this chapter, we looked at the standard definition of Industry 4.0. It is a visionary statement, and there are many companies along the path to achieving that vision. However, many companies are just getting started or thinking about how to get started. Industry 4.0 is about data and the management of that data. Alongside data comes the necessary analysis, information, knowledge, and the innate ability to improve by looking at the right things. Industry 4.0 allows us to go beyond the decades-long approach of the status quo. We know from history that at every phase of change, in probably every sector, many felt that change was not required or too fast.

How IoT can support Industry 4.0 at scale

The authors of this book have spent a lot of time working in IoT, going back well over 10-12 years from when IoT was little more than a buzzword. When we think about IoT, our minds go to cheap, easy-to-use hardware and connected appliances or watches. This new crop of inexpensive hardware has opened people's eyes to what could be done for minimal cost, but for industry, a different level of hardware is often required.

We can use IoT hardware and software to accomplish the goals of Industry 4.0 by providing a robust and industrial-strength set of technologies that allow for the instrumentation and measurement of equipment and its environment. Bear in mind that industry is often conducted in extreme environmental conditions and the cheapest approach is often not the right approach. A trade-off between cost and reliability should be considered since if you have to replace a component too often, then the value can be lost in effort, time, or the loss of data while waiting for the switch to occur.

Let's talk about the key areas to be considered when turning to IoT as part of the solution. Let's say we are going to place a simple sensor on or near a device to measure temperature. We don't need to be specific at the moment, but just consider the conditions that you might be facing, such as the following:

- **Tough**: Can your sensors and equipment withstand environmental conditions and pressures? Industrial equipment in the field can be in a rough environment. Does the sensor and corresponding transmitter require an **ingress protection** (**IP**) or **National Electronic Manufacturers Association** (**NEMA**) enclosure rating for protection? IP ratings provide a rating for your enclosure for protection against access to the internal components and protection from the ingress of liquids and dust or dirt, which is essential for harsh outdoor environments. An IP67 rating indicates a solid enclosure that is protected from the ingress of dust and protects against temporary immersion in water up to a few centimeters. NEMA ratings are the same as IP ratings but provide additional classifications against corrosion and hazardous locations. For some environments, such as oil and gas, a NEMA Class I or Class II enclosure is required due to the presence of corrosive liquids, flammable gasses or vapors, or combustible dust. These environmental conditions and requirements can add additional costs and time to your effort in sourcing, testing, and possibly certifying your components for use in the field.

- **Easy to deploy (and maintain)**: Make it as simple as possible to ensure speed and accuracy when deploying equipment. This ensures that deploying and registering your sensors and equipment is simple, almost bulletproof, for the engineer on site. When deploying a sensor to a piece of equipment or a location, we have to ensure that once the sensor is in place and operating, we can tie it back to the right location. Without that, the effort is useless, and none of the data further up the chain will be reliable. There are several options here. Mobile apps with barcodes and even manual configuration are fine as long as the setup can be done correctly and consistently. Additionally, the sensor should be easy to attach and place. OK, simple is not always possible, but as much pre-configuration as possible should be considered, leaving the engineer to do as little as possible on-site to complete the setup and installation. Runbooks should be well-defined and include any troubleshooting information that might be needed in the field. This is especially true if the deployment people are not experienced in the new technology.

- **Scalable**: The ability to quickly deploy many sensors in the field should be considered. This can mean dozens, hundreds, or thousands of sensors across multiple locations or across the globe. Both hardware and software can be a concern when thinking about scalability. Something easy to deploy and configure can be deployed by the thousands; however, if the software or storage is not configured to manage the data, it may result in wasted effort. Cloud technology will help with the software part, although the application requirements to view and analyze data need to be able to keep up as the system grows. This means data systems and analysis should be designed to accommodate the potential millions of readings you might expect from all those sensors.

- **Reliable**: This ensures sensors and monitoring will keep working over a long period of time. This is not the same as a sensor or node being tough. It's about reliability. Reliability is much more important because now we are talking about the electronics, rather than the casing and packaging of the node. Do the electronics have sealed or glued connections? Are the sensors potted or otherwise protected? Potting means filling or surrounding the electronics with some type of gel or epoxy resin, essentially encapsulating the electronics to minimize the dust, vibration, or liquids that affect them. Of course, before going to this extreme, quality control and using high-quality components are recommended. Hot glue on your connections can be the first line of defense against the loosening of wires. If you go to the extreme of potting your components, be aware that it cannot be undone, so when a part goes bad, a replacement for the entire assembly will be needed, which may be costly. Carry out a cost-benefit analysis of the best approach based on your industry to make the right design decisions.

- **Secure**: Make sure data and systems are protected from malicious actors or data theft by unauthorized parties. We will talk about security throughout this book. Using IoT technology can potentially leave security holes across the entire data stream. There are several aspects here to consider. Ensure that data is secure while traveling upstream, and protect endpoints so that fake data cannot be introduced into the system to influence results or actions. Since we are talking about the endpoints in sensors or nodes, physical security is the first step to consider here. Do you need to apply any physical security to the deployment location? And if not, what type of tamper monitoring can be put in place to alert you if something seems amiss? Once data hits the cloud, traditional IT security methods can come into play, but with equipment in the field, your system can encounter many different types of threats.

If you are in the industry already, you will already know some of this; if you are an operator, you will know your environment intimately. But it is still worth considering the environment and defining some basic requirements alongside your goals. This is an excellent point to think about the domain you are considering and create a simple checklist with critical criteria that you can build on as we go forward. These considerations go hand in hand with each other and drive toward a common goal. Using the preceding criteria, along with more to be added as we go, can help you navigate decisions and communicate to others the expectations of the technology.

Sensor technology

We use the term **sensor technology** broadly to include both the sensor or sensors involved in instrumenting an environment, but also the **sensor node** that reads the data from the sensor and transmits it to the receiver. We separate these currently because they do not always go hand in hand. Sensor technology is constantly evolving and transmission protocols, such as **low-power wide-area network** (**LPWAN**) and 5G, continue to evolve. The sensors you wish to use may have specific characteristics and may not always be compatible with a sensor node for transmitting data on the protocol that you have defined. Let me share an example.

Several years ago, during the Wild West of IoT evolution (just kidding, it's still the Wild West), one of the authors was involved with a proposal for a large city in Nevada; you can probably guess which city. Our partner in the deal was an IoT start-up; actually, it was way more than a start-up, with millions in funding, but it was relatively new to the IoT space. The company had some fascinating communication technology, which should have been a strong competitor to cellular technology and most LPWAN technologies. It was low power and could send data over very long distances, outdistancing other technologies such as LoRa (from long range) by miles.

This company spent a lot of money and energy on trying to sell and further develop its network, and since fewer towers or hotspots were required to blanket an area, they felt they could cover large areas, such as a city, with relative ease and lower cost. While this was probably true, little regard was given to the fact that no sensors or sensor nodes were available to use on the network. Chips were available and provided by the network provider, but the cost, time, and effort were left to sensor vendors to implement. Essentially the sensor vendor or consultant had to make a bet on this working based on little more than faith in the network company. In hindsight, it's a bet we are glad we didn't take.

Unfortunately, this turned out to be a failed strategy since the investment was too significant and complex compared to more available options at the time, using protocols such as LoRa, Sigfox, and LTE. I'm still disappointed the company didn't have the vision to see this hole in their strategy, and they have moved into the realm of also-rans in the IoT space.

The key takeaway here is to keep in mind the following:

- Can your sensor node or transmission unit communicate back to the cloud with your chosen protocol, or set of protocols if you need redundancy?
- Can your sensor node communicate with your actual sensor or set of sensors to read the measurements for data transmission?

There can often be a mismatch here as the sensors themselves can use all kinds of unique protocols. For example, *SDI-12* is a standard serial communications protocol used in agriculture and weather sensors and can be challenging to read if the sensor node is not designed for it. The protocol was defined in the late 80s and transmitted ASCII characters over a single data connection. There are many examples of serial protocols in place for industrial systems that can be decades old but are still very much the standard.

Another example is if you need calibrated sensors, such as temperature sensors, that must follow *NIST* standards to ensure the results adhere to standards. If calibrated temperature sensors using your defined communication protocols are not commercially available, then you have limited options.

Every day it seems, the sensor world gets a little bit brighter as a vast array of sensors, edge devices, and transmitter units or nodes become more readily available. Many of the most popular protocols are available, with new ones coming on slowly as new network technology is better adopted by the community and becomes more readily available. However, there are still sensor solution gaps for many situations.

One of our favorite options for this problem is from a company called *Libelium* out of Zaragoza, Spain. Libelium offers a robust mix-and-match approach to sensors and communication options of all different types. For example, you can choose sensors for measuring air quality, water quality, security, and agriculture or for integration into industrial protocols such as *Modbus*. You can pick a communication protocol to connect the sensors and send measurement data to an existing application or web service. Protocols include using anything from LoRa to Wi-Fi to 4G. This flexible approach makes it easy when you try to adhere to a standard communication protocol but cannot find an appropriate sensor that works with your chosen standard.

Cost can certainly be a factor, especially at scale, and while prices seem to be continuously going down, here is where the myth of IoT, again, seems to get in the way of Industry 4.0. You get what you pay for, and this can be crucial in harsh environmental conditions and areas where you need to provide a standard approach.

IT versus OT

There is still a lot of confusion around **information technology (IT)**, **operational technology (OT)**, and this idea of convergence. But essentially, it is a simple concept to understand.

IT is something we are all familiar with in our daily lives. We run applications on our phones or laptops. Many of these applications run on servers or in the cloud and process data-producing orders, sales, and directives or provide some type of analysis. This is the IT world that we know today. It's a reasonably open world, and access can be gained from anywhere (provided security concerns are followed).

OT, especially legacy systems, can be considered a more closed environment. OT ends within the walls of the factory. When you think of OT, think of **supervisory control and data acquisition (SCADA)**, which is also run by servers but interacts with devices within a defined area of control. At a large scale, consider a power plant or water treatment plant. The pump shuts off when the water in a tank gets too high. Too low, and it starts back up again. Monitors and alerts allow operators to visualize and help manage what is taking place with appropriate alarms and controls.

Industry 4.0 and organizational alignment

Figure 1.1 illustrates how different areas within the business fit together and into the big picture. In order to work, there is a strong dependency across IT, operations, and business and management; all stakeholders must work together to realize the benefit.

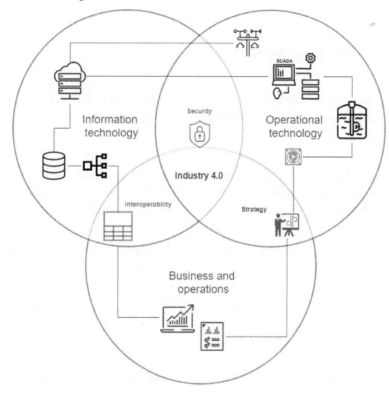

Figure 1.1 – Industry 4.0 organizational alignment

OK, what is the big deal? The big deal is that IT's primary goal is to provide the business and management with information and the ability to support the decisions and operations of the company. How many widgets were produced? Or how many barrels were processed? How many were sold, and at what price? The business lives and dies on this information and data being available faster and more accurately to provide a competitive edge. Often what is missing is an insight into the real-time production of widgets or processing of barrels. Newly built or upgraded factories can provide real-time information, but in legacy systems, even relatively young ones, that information is hidden. And in production, modifying (reverse engineering) devices and machines voids warranty and, if not done correctly, may lead to complications. With the emergence of IoT, we can bring some of that data from the closed OT world into the often more integrated IT world, where it can be used more effectively.

The focus of this book is on getting the hidden data, storing and processing it, and then using this information effectively.

The business is not the only one to benefit from introducing new data. Operational teams will gain insight into the equipment and production that they didn't have before. Uptime and maintenance can be improved, cost reduced, and throughput increased as a new understanding, and a new normal of the environment begins to emerge. The full benefit of digitalization should become clear in the rest of this chapter and throughout this book as we share examples of collecting data and then using that data to realize value across the organizational spectrum.

You can get there from here

Industry 4.0 is driven by IoT, but it is just one part of the picture. A big part, granted, as it allows visibility into equipment and operations as never before. A longer roadmap is required to achieve the vision of digitizing your industry and the transformative changes that can take place.

> **Important note**
>
> We are not a fan of big, complicated, eye-chart-type visuals, so throughout this book, we will keep the visuals simple and coherent to allow you, the reader, to immediately understand the concept rather than asking you to try and understand something overly complex.

Figure 1.2 illustrates a basic roadmap toward digitizing your industry or moving toward Industry 4.0. We have broken this down into four primary areas of consideration for improvement. Within each of these areas, there is a vast number of considerations both on the technical and business side to consider.

For many, the status quo or current state of their process is *operate*. Consider this business as usual, and maybe decades-old processes that, for the most part, just work. There may be some instrumentation, perhaps even a lot, but no cohesion or integration across machines, systems, or plants. Everyone knows we can do better, but how do we move forward? *Figure 1.2* illustrates a set of steps for continuous improvement in your equipment and environment. Instrumentation and acting on it improves both the business and technical responses of the organization.

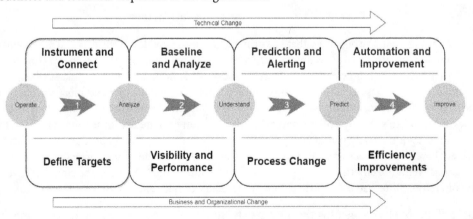

Figure 1.2 – Industry 4.0 roadmap

Let's talk about each step of the process outlined in turn. We have labeled each area based on the technical changes because that is the focus of this book. However, this can be adapted based on your principal needs.

Instrument and connect

Moving beyond the general *operate* state requires in-depth visibility of your systems and environment. Consider this the instrument phase, where the goal is to start to gather data from your systems and environment. The other side of the equation is to know what should be instrumented and why. This effort of instrumentation and collecting measurements is where business and operations can collaborate to ensure that the data collected is needed and understand how that data will be used to drive processes and the business forward.

It is usually not the best strategy to jump in and instrument everything. While it may seem like more is better and that you have nothing to lose by doing so, spending time and money on equipment, manpower, bandwidth, and storage for data that is never used ends up being a losing proposition. Once committed, it may require ongoing maintenance for data that does not provide good value.

Another question to ask is how much data is needed. This depends on the velocity of what you are measuring and collecting. Some systems can churn out hundreds of measurements per second. How and where should this information be stored and analyzed? Does all of it need to go to the cloud? Can we process this on the edge and provide aggregate results? What are the pros and cons of taking an approach toward managing this data? Business and operations should be involved intimately in these discussions to help drive what level of granularity is needed and how it will be used. This in turn can drive IT decisions for data management and processing.

Baseline and analyze

Baselining your system's normal operating environment can be an eye-opening experience. Sometimes (actually a lot of the time) we don't really know what *normal* is for our equipment until we measure it and then see it in some graphical format. SCADA systems often have this insight into pressure, temperature, and flow characteristics, but not all industrial operations are driven by SCADA, or the information is hidden from all but on-site equipment operators. The insight gained here can be enormous. Measuring a handful of values can provide deeper information about the working condition of a piece of equipment or an end-to-end system and, as we will see later, drive efficiency and potential maintenance issues. Understanding the baseline of system performance and conditions at a known production rate can be powerful, as well as asking questions such as, what happens when the production rate goes up, and how does that affect the machine conditions?

Defining a baseline can take a long time; it is not done in a day or even a week. Expect at least several production cycles, which could be seasonal-based activities that could take months or years. Hopefully, most cycles do not take that long, but if your industry is influenced by weather or environmental conditions, there is that possibility. You can continuously gain good insight by getting comfortable with what your baseline looks like along the way, but unexpected curves and influences only occur with time.

Prediction and alerting

From a technical side, we have many opportunities. Now that systems are being more closely monitored, you should expect to see variations in the data as problems occur, and equipment shuts down. Maybe there were some unexpected vibrations or a temperature rise before it occurred. Can we monitor for a particular set of variances? Do the vibrations occur when a part is ready for replacement or maintenance? This is the beginning of condition-based maintenance, where new data or real-time monitoring of the environment can alert the operator to a possible set of conditions that may fail.

To accomplish this, we need to start to build predictive models. Tooling today can make it a relatively straightforward process to create a predictive model; however, much of the work in the baseline phase will help you determine which data to prepare for modeling. Generally, we are looking at data to help predict downtime or failures of equipment or systems; however, does the data do this? We will dive into some details about predictive modeling and how to use this in your architecture in the coming chapters.

We can often start our journey by using simple thresholds or comparisons on specific values or sets. This is especially true when you know what specific events or conditions you are looking for but are not quite sure how predictive models will advance your cause. Does the temperature rise above a specific degree? Does the energy usage on a pump get higher while the pressure gets lower? These are simple examples, granted, but powerful tools in helping to determine when something might need to be checked. At this point, we are still triggering more manual alerts, effectively telling someone to check something. This could be as simple as an email or SMS, or a more advanced trouble ticket being opened automatically on your **enterprise asset management system**.

But really, we can now take this further into the business side of things and better understand production cycles and issues and capacity constraints, not only of finished products but subprocesses that may cause bottlenecks.

Automation and improvement

Industry 2.0 and 3.0 brought a lot of automation into manufacturing and processing. Our focus is more on the automation of the overall business and what is produced. The ability to monitor and eventually steer your production closer to real-time allows the business to be more agile and respond more easily to customer demands. This is a topic well beyond the scope of this book, and would possibly include connecting customer demand, supply, and fulfillment, as well as the digitalized production or factory that is our focus here.

However, with a deeper analysis of your historical data over time, a more detailed analysis can occur of where and when improvements can be made.

Visibility is everything

We probably can't say this enough. Possibly, this is the gist of the entire book, along with some focus on what to do after you have better visibility. It was mentioned before that understanding your baseline, or *the normal operating conditions of your environment* can provide clear insight into what is truly normal and when some type of abnormality occurs. This can only happen with clear instrumentation. This is true in almost any industry or science. Most experts will explain that the instrumentation of your environment allows you to gain new insight with a precision not previously available. Software developers who have used deeper inspection, such as bytecode instrumentation or injection, can easily explain the advantage of increased visibility into aspects of a running system. The same is true for physical systems and being able to view and analyze the physical characteristics of a piece of equipment or an environment.

Another aspect to consider is global or widely distributed operations. Modern equipment or systems can be outfitted with all the instrumentation needed for the safe and efficient running of the process. However, what about systems that are geographically distributed across vast areas? Combining and even comparing information from systems globally can provide new opportunities for decision-making.

Along this path should be a feedback loop, allowing adjustments and updates across the entire monitoring chain.

Business driving innovation

A quick web search will provide an abundance of IoT information, specific lists of ideas, examples, or use cases for implementing IoT for your industry, and the value it provides. Sometimes there are interesting use cases, but often it is driven by marketing and sales looking to sell their solution. Unless you have a good working knowledge of an industry, this can be misleading. Earlier, we mentioned that just because you can instrument something does not always mean you should. Time and cost considerations should come into account. Consider the cost of adding sensors to gather information, but also the data collection and maintenance costs of continuing to gather data.

IT, operations, and business stakeholders need to work together to understand what it is that they want to achieve. Then real subject matter experts need to be involved in telling you how to get it and then interpret the right data points to achieve those goals. Operations understand better than anyone how to develop, manufacture, or produce materials or goods. Business stakeholders know how to price, sell, and distribute those goods to end customers. There are nuances in business and operations that the other may not understand intimately or agree with, but working together to achieve better visibility and control can be a powerful weapon for competing on the global market.

The truth is, business and management may not know what they need to instrument at a detailed level. But they do know what information they need to make decisions, such as better **overall equipment effectiveness** (**OEE**), downtime reports, or more detailed forecasting for service lines over a period of time. OEE is a process for measuring manufacturing productivity by looking at equipment availability and performance, and the quality of manufacturing output. Operations can then make an informed decision about what they need to do to obtain and provide that information. It's a complex process that is greatly oversimplified here, but hopefully provides some guidance that no one area of the business should work in a vacuum on this endeavor.

So far, we have provided a big-picture overview of Industry 4.0, the digitalization of the industry, and Industrial IoT. There are multiple approaches to accomplishing systematic improvement in your production and management of equipment and processes, and the roadmap is one approach. Moving forward, we want to dig deeper into some of the technical aspects of starting your journey and adopting a digitalization mindset and approach. What are the steps and goals for moving forward and getting incremental value along the way? In addition, what are the pitfalls in adoption and understanding how difficult it will be? We will be exploring more of the idea of instrumentation, analysis, and convergence for providing value across all stakeholders.

The convergence – IT, OT, and management working together

Rarely does an opportunity come around in the industry for all aspects of the organization to come together for everyone's benefit. Industry 4.0 allows that to happen. The digitalization of industry can benefit all aspects, allowing the business to make better decisions around schedule, price, and volume and providing operations with better tools to make decisions about maintenance, downtime, and upgrades of equipment within a plant or factory.

Evolving toward IoT and Industry 4.0 allows for many things to occur. Maintenance cycles can be improved, which positively affects planning and production. The roadmap toward the continuous improvement in your maintenance approach outlined in *Figure 1.3* provides an idea of how your outlook and planning can be improved over time.

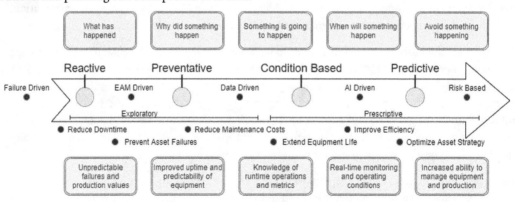

Figure 1.3 – Driving toward common operational goals

This is not an overnight process to achieve results with instrumentation and monitoring improvements, so we wanted to talk about the progression and the maturity curve that might be adopted moving forward.

Reactive and preventative maintenance

Many companies are in the first two categories of reactive or preventative maintenance.

Reactive maintenance is where we basically run our equipment until it breaks. When a piece of equipment or production line breaks, or when we notice some issues with the machine, then we fix it. For example, we replace the belt when it breaks. This is fine in some cases where the fix is simple and relatively inexpensive, but if we need to order parts or the replacement effort is extensive, it could result in unexpected downtime and poor production results if the break is in the middle of an important production cycle.

Preventative maintenance is a little better. It's often driven by the calendar or perhaps usage-based (number of hours), similar to changing the oil in your car every few thousand miles. It may be driven by an enterprise asset management tool that is put in place to manage inventory and track and manage assets based on tickets from the field or by using the provided manufacturer guidelines for scheduling maintenance. So, for our previous example, the belt is scheduled to be replaced every few thousand hours based on the manufacturer's recommendation.

The advantage of preventative maintenance is that downtime can be better scheduled, and unexpected downtime can be reduced. It may be the best you can achieve with in-depth instrumentation; however, it may not be enough. One concern is that you may over-maintain your assets or under-maintain them, which can cost you more in maintenance or asset replacement costs.

Even as we move further down the optimization chain, I don't think we will ever really replace these types of maintenance. Things happen, and reactive maintenance is necessary. And yes, that belt should be replaced on the recommended schedule. Production machinery almost always needs constant love and care. But the long-term benefit is that you don't have to devote attention to it as if you were in the Army, where if you are not training or sleeping, you are performing preventative maintenance on your vehicles, weapons, whatever, whether it is needed or not. Of course, in this case, the ultimate goal is to keep you alive, so not a bad thing in context.

Condition-based maintenance

As we start to initialize data and collect more information from the individual assets, we can move to condition-based maintenance. We can look closer at the real-time conditions around that asset. Measurements such as current output combined with solar conditions or compared with other solar modules can help us understand whether a component is underperforming and may need to be looked at more closely.

With **condition-based maintenance**, we get into the habit of not performing maintenance too early or too late but in line with actual data from the equipment and its performance.

Predictive maintenance

Predictive or **prescriptive maintenance** takes this one step further by adding **machine learning modeling** to remove the human component from the monitoring process and provide feedback based on what it has learned about optimum operating conditions. Some would argue that these are two different things.

Predictive maintenance helps to forecast potential problems or outcomes with your environment. Prescriptive maintenance helps to provide recommendations based on those outcomes. Neither of these is new to us. Consider the wayfinding application on your phone, which tells you in advance of a delay ahead on your route. Many of these applications will prescribe alternate routes and determine the time savings or delay.

Eventually, some companies may lean into **risk-based maintenance**, which helps you consider the risk that this part will fail. How will production be affected? Or what downtime may occur if it does fail? Essentially, what is my cost if I keep running and a failure does occur?

Approach with caution

We have all seen or heard of the *don't-touch-it* mentality from production teams who just need to keep their system running. When systems are old, decades old, there can be a lot of lost knowledge about how those systems work. Parts may be hard to replace, and sometimes, things work that shouldn't. These are some of the challenges that operations teams face, and it can be hard for anyone outside those teams who need to get a better look at what is happening inside.

These types of environments and equipment need to be approached with an abundance of caution. Trust is a big factor, and sometimes IT needs to earn trust and not come in like a knight on a white horse, suggesting it will fix all the problems of the day. Everyone in the organization needs and wants progress, from the boardroom all the way to the equipment operator on the shop floor. However, moving too fast in some challenging environments is the easiest way to lose trust within the organization.

We have found from experience in many different industries that most operators love to share information about how they do their job, how they treat their equipment, and, mostly, how things could be made better. Having worked in consulting for many years, it is incredible what you can learn when you talk to subject matter experts in the field, learning everything from farming to fracking, with experts always being generous with their time when approached respectfully. Here are a few simple steps as an approach to this. Some of these tactics seem almost silly to say out loud, but they're essential to keep in mind:

- **Listen and learn:** Understand what all the stakeholders know about a system and how it works. What are their concerns, big and small, and what suggestions can they provide for improvement moving forward?

- **Define success:** Understand what success means, whether this is for one piece of equipment, an entire line, or the plant. Knowing your end goal can help you stay focused and on track.

- **Research and share:** Learn more about the options you may have for instrumentation and sensor deployment. Share what you have learned with the operators and discuss the pros and cons. Do not get fixated on a single path until you find some general consensus on what could work. Even if you know you are right, it won't hurt to bring everyone along with you through consensus building.

- **Proceed with caution**: Finger-pointing when something goes wrong can be a disaster, not only for the people involved but for the project as a whole.

Sending up trial balloons (or conducting a spike in the agile world) for testing can be a viable approach. Testing out different sensors or interfaces without full commitment will allow you to evaluate the pros and cons of the approach and share the results with stakeholders.

Of course, all these guidelines are focused on production equipment, when a mistake is liable to cause serious problems through equipment downtime or worse. A standalone temperature sensor doesn't require nearly as much effort, but it's probably not wise to grab the first one you see and run with it. Even low-impact sensors should be evaluated against some of the guidelines discussed earlier. We outlined some considerations when sourcing equipment and sensors, but we have not addressed the approach for determining whether the sensor can provide the correct measurements. Ensure that you discuss equipment tolerances, temperature fluctuations, and pressure and vibration values. This allows you to choose sensors that can operate within the machine tolerances and provide measurements at the right frequency.

Leveraging good architecture to drive progress

Architects are big fans of patterns and standards. Doing something in a similar and proven way is common in just about every type of job or industry, especially those that produce physical or tangible results. The size or strength of the material can determine much about how to build something, and it helps to define the time and cost that goes into the build. We can translate this directly into building something less tangible, such as software. With Industry 4.0, we have the best of both worlds, hardware, and software, allowing us to use traditional engineering techniques and hopefully rigid hardware and software engineering processes.

In software, we often call them best practices, but it is not always everyone's best practice. Sometimes what works for one company or team may not work for another. Or similarly, differences occur across industries or environments. With this in mind, our goal is not to prescribe exactly how things work but rather to provide a set of strawman models based on generally accepted industry best practices that can be adapted to your situation or solution.

Throughout this book, we will share approaches and patterns that can be adopted but don't be afraid to go your own way (within reason). In addition, technology changes over time. What might be the best practice today may not be the same tomorrow as new approaches evolve for achieving your goals.

Observability

We have mentioned the idea of observability in this chapter, but let's review it in more detail. Observability is not quite the same as monitoring, but there are a lot of different opinions on what the difference is. The name is not that important, but conceptually you should understand the difference. Our best definition to build upon monitoring techniques with an observational approach is as follows.

Monitoring involves collecting data from sensors, logs, and production or machine metrics. This allows you to understand the current state of a system, whether that be a machine or even the components within your IoT application. Monitoring enables you to know whether something is working and when a system or piece of equipment is down.

Observation takes it a bit further and allows you to better know why something is down or not working at the optimum level. Observation requires monitoring and possibly additional instrumentation at multiple levels. Consider the following possibilities:

- **To see how systems work together**: In IT, it is often possible that one system or process affects another, but clearing away the potential issues to get to the root of the problem can be challenging. For example, a web page is loading slowly, but the problem is that the database is overloaded or that the query is more complex and takes additional time to complete.

- **To gain a deeper understanding of an internal systems operation**: In the previous example about the database, only through deeper instrumentation, such as bytecode injection in the code world, could we gain insight into what is happening. That instrumentation level also allows us to see how components work together and understand how one depends upon another in the process flow.

To summarize, we can distinctly define the difference between monitoring and observation:

- *Instrumentation supports monitoring, which tells you that something is wrong*

- *Monitoring supports observability, which tells you why something is wrong*

As I mentioned, there are different interpretations (I have seen examples where this is reversed), but this seems to make the most sense. In reality, the end goal is something we all want to achieve. We want complete visibility into our systems and processes so we can handle problems and improve the outcome.

Repeatability lowers cost

As you move forward toward advanced digitalization, there will be a lot of moving parts. Both hardware and software will need to constantly adapt as the solution evolves. For large implementations where there is a lot of equipment, or the environment is widely distributed, it may take a lot of help to deploy and manage equipment. Items to consider, include the following:

- What type of equipment are you using, and can it be deployed or configured to handle multiple uses or types of equipment?

- What is the best purpose-fit type of sensor monitoring equipment (single-purpose or multi-purpose equipment)?

- How does sensor monitoring equipment stand up in terms of cost, training, installation, and maintenance?

- How much field training will be needed? Simple plug and place, or complex configuration and management training and runbooks?

- How will software need to be adapted as sensors and monitors are deployed?

As your solutions evolve, consider these guidelines and how you can streamline the deployment and management of the solution, ensuring that as much repeatability as possible is followed. If sensors and data collectors need to be configured, do the software and cloud need custom work? Think about things such as custom tags. At scale, how much work will it be to collect and store data from hundreds of pieces of equipment and thousands of tags?

Summary

The overall focus of this book is implementing Industrial IoT and leveraging AWS cloud technology to achieve the goal of digitalization of your industry and environment. This chapter was designed to set the stage and provide a basic understanding of Industry 4.0, including its purpose and a view into the road ahead, helping you to have a better understanding of the end vision. A strong roadmap and vision, guided by business goals and coordinated with operations and production teams, can help you stay on track and ensure that everyone buys into the process.

In the coming chapters, we will focus on more technical details of how to achieve some of those goals toward instrumenting, collecting, and using data within your environment. This is the fun part of the process for IT geeks who are interested in learning and using new technology. The advent of IoT has opened the availability of hardware to software engineers who don't always have the opportunity to work with both. Hardware, software, networking, and analytics are now considered the full stack of most IoT engineers and open up new opportunities for interesting solutions in the industrial space.

In the next chapter, we will outline our overall architecture planning approach. We will describe the main architecture levels and key components to build a successful and scalable Industrial IoT platform. Starting with the basics will provide a starting point for adapting some useful architecture concepts to your environment as you initiate or continue your IoT journey. We will also talk about the roles of the IT architect and different types of architect roles and discuss how you can add the most benefit to the project's success.

For now, grab a cup of coffee, sit in your favorite chair, and enjoy!

2

Anatomy of an IoT Architecture

As we start to think about the architecture and set of solutions, it is essential to have some idea of where we are headed. In *Chapter 1, Welcome to the IoT Revolution*, we reviewed the concept of making sure that business and production teams are aligned with their goals. However, in this chapter, we will focus more on the IT and IoT side of things and how we accomplish those goals from an architecture perspective – how data should be collected, how it flows and is managed through the system, as well as where data starts to be processed, and what stages of processing need to be considered. And then, finally, we'll learn how to make that data and the resulting analysis and information available to the business for decision-making.

These decisions are based on multiple factors, such as your industry, business goals, availability of gathered data, and the type, location, and velocity of data you are collecting. This chapter hopes to provide some initial steering in your approach and decision-making, guiding you through options and choices. This book has focused on using **Amazon Web Services** (**AWS**) as the cloud service provider for all the examples; however, in this chapter, we will continue to focus on concepts, for the most part, that can be adapted to several cloud solutions. In this chapter, we are going to cover the following main topics:

- Architectural thinking
- Architectural components for IoT
- Bringing it all together
- Defining standards

As we progress through the architectural layers, we will identify several AWS products and services that can be applied within different layers. This will help you better understand how different services can be applied. These are just initial examples of different types of services that are available without concentrating on whether or not they are the right services to choose. In later chapters, we will start to look deeper within the layers and make concrete choices about which services to use for our specific solution.

Technical requirements

A good understanding of cloud technologies and services will be helpful in this chapter. Although we are looking at things more generically, one of our goals in this book is to provide design and implementation advice on building solutions using AWS. Therefore, some general knowledge of AWS and some of the IoT services will be helpful toward your understanding but are not necessary. If you are familiar with another cloud provider, you can probably map those services directly into the layers we present.

Architectural thinking

Building a great IoT architecture requires a good understanding of the steps involved and how everything fits together. Most of us know the basics, at least conceptually, of how to collect data in a variety of ways and then move data into the cloud but putting together an end-to-end flow that allows you to process, store, and visualize data and subsequent analysis is the challenge. There are many ways to accomplish these goals, and not thinking it through could have consequences. This could be costly if a rework is required in the future. On the other hand, too much analysis and overworking of your architecture can also have the effect of costing you more time and effort, both initially and over the long term.

Architects often need to walk a very thin line between the traditional approach of designing for every occurrence (what if we need this function or that feature in the future?), and a more agile approach of building just what you need, when you need it. I often tend toward the latter approach, especially when I know that time and budget are not in our favor. Everyone loves to exceed expectations, but this can be problematic if you risk not meeting any expectations as a result of trying.

Reference models

In today's world, generally, we can start with a reference model. Reference models are a wonderful thing in IoT, or really in any technical or architectural endeavor. It wasn't too long ago that we were figuring things out as we went, literally building the reference models for the future. Reference models provide benefits in several areas.

First, reference models guide us on how things can fit together. Most reference models are based on best practices taken from other teams or cloud providers who have been there; done that! As a starting point, this can remove some of the initial concerns around if you are doing it the right way. It is never one-size-fits-all in this field, so adapting a model or approach to your environment is probably necessary. You may prefer a different cloud provider or have a unique environment that needs to be heavily customized. Adapting and building your interpretation is perfectly understandable and expected.

Second, reference models provide a way to communicate across the organization where you are headed. Depending on the level of detail, this will ensure that all stakeholders understand and agree on the approach, that the implementation team knows how to move forward, and that there is consistency across the organization as your Industry 4.0 plan moves forward.

When building a reference model, it is important to be cognizant of designing overly complex models or diagrams. Sometimes, this is necessary, but often, it turns into an eye-chart experience, where few stakeholders grasp all the complexity. You should know this is a pet peeve of ours. A better approach is to divide the model into different levels of complexity, starting with a high-level 0 type of chart and then drilling down with additional models for 1, 2, and 3 levels of detail.

Often, reference models are used to communicate to the business and technical team about how things should work. With complex systems or environments, we can start with levels to illustrate concepts simply. This allows non-technical stakeholders to grasp your approach, without having to wade through obtuse technical details that may not be of interest or wholly understood by non-technical readers.

If you are starting completely from scratch and just trying to get an idea of how things might look, consider best-of-breed products, and best practices from some leading providers. It can be tricky since there is a strong bias about being locked into some vendors' plans to use only their products. But it can also provide the necessary starting point for understanding how some of the latest technologies can work in your environment. Take it slow and steady, until you can wade through the noise and adopt the right technology for your goals.

Architectural components for IoT

To understand where we are going, we sometimes have to know where we have been. The history of IT and industrial automation can help with understanding how things work today, and we can build on them as technology evolves. One of those earlier reference models is the Purdue Reference Architecture developed in the mid-1990s. One of the key characteristics of the Purdue model is the definition of layers or levels of separation for industrial systems.

ISA-95 is the standard for Enterprise-Control System Automation, which was built on top of the Purdue model and continues to evolve today. The focus of ISA-95 is to help define standards for interfacing production control systems and enterprise IT systems, focusing on layers 3 and 4 of the Purdue reference model. This is generally where operational protocols such as fieldbus protocols meet the IT or Ethernet networks.

In large part, we are moving into the realm of **Enterprise Architecture** (**EA**) and defining the architecture, standards, and vision across the entire organization. But we must also merge that vision with operational dependencies, restrictions, and goals. One of the defining goals of EA is to install discipline within the architecture while at the same time providing the capabilities that the organization needs to operate and evolve.

EA is a large topic and is now split between technical and business architecture, which is a lot to consider. This book will look at the big picture in many areas, but also focus more on the hands-on implementation of some of the technology discussed. However, the lessons learned from this book can be applied to your EA's due diligence in determining the best way forward for your organization. The patterns that evolve from your effort here can be enabled across the enterprise as future standards.

Designing in layers

The Purdue reference architecture defines six levels that are split into several zones. This is a great start for us to understand digital automation and the integration of industrial systems, and we want to provide a sense of familiarity for those already comfortable with these levels. But technology evolves at a rapid pace, and our technology stack more than likely has greater capability than previously available. So, to adapt as necessary to accommodate technological changes (the technology crank is always turning) over time, our stack will look different.

Our approach is actually to view our model more from the standpoint of IoT and cloud architecture. We are focused on the cloud, we must get data to the cloud and store, analyze, and present it as efficiently and robustly as possible. This allows us to take full advantage of the latest best practices and architectural viewpoints for this technology. We can then layer in industrial controls and systems as appropriate for our goals:

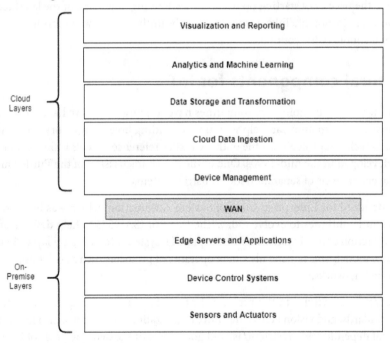

Figure 2.1 – Layered architecture outline

Figure 2.1 is our first glance at a reference layered architecture. I have avoided giving these layers specific numbers since I think there should be some fluidity based on your circumstances. However, this model can be easily mapped to the Purdue model, or other industrial models, or modified completely based on your experience and preferences.

There is also the chance that, with many digitization efforts, some early models will not fit your vision. The stricter communication path from sensor to end user can break down as new types of communication protocols and sensors evolve that can go directly from the sensor to the cloud. Our goal is to keep this architecture model flexible and allow you to include additional considerations in defining the architecture, allowing you to make sure your architecture adheres to organization goals and standards, not someone else's vision.

Large organizations with many facilities may face additional challenges, especially where there is little standardization and commonality across factories. Some of these challenges are well beyond the scope of this book. Hopefully, we can provide some guidance in leveraging the cloud on your journey, but injecting structure and applying EA guidance within any large organization requires significant effort.

Consultants often refer to processes as being a significant driver of change, and this is very accurate. The best technical solution can fail if internal or business processes don't change. It is difficult for outside forces to influence this change within the organization. It has to come from within the organization, and our responsibility is to show why a change or process restructure will be better for everyone. Communicating the value to the organization and the real benefit to individual stakeholders is the primary challenge in this process.

Defining the layers

The layers in *Figure 2.1* are a strawman representation of what you might consider your initial design attempts, which is the focus of this chapter. Adding some definition to each layer will help us define how we might use these layers and draw distinct borders between these layers. When thinking about borders across layers, it is important to consider communication options, protocols, and security as data crosses the boundaries that are defined.

While we will spend a bit of time defining each of the layers, our ultimate goal is to use them in our architecture. One question is *why*? Why do we present and suggest that you follow a layered architecture? There are many benefits to this approach and the precedence goes back decades, from initial multi-tiered distributed computing architecture approaches to current cloud best practices.

Having layers allows you to group similar functions, such as data ingestion, storage, or analysis, into distinct areas. This allows your processing logic or business logic to reside together. Having layers also support the idea of a separation of concerns. That part is pretty obvious, but it should be stated directly that data ingestion, storage, or perhaps data transformation should occur in one layer. This will allow changes in specific functionality to occur more easily if these functions are contained in one area. For example, decoding data messages can often change as new measurements are added, or a different sensor is deployed with a specific message type. Confining those changes to a specific layer and set of services helps with managing the change across the environment.

Another consideration is that most of the data collection, transformation, and analysis can happen in more or less real time. In many cases, this can be clarified as "near" real-time analysis since the expectation is not instantaneous. A delay of several minutes or more should be expected as data winds its way from the machine through processing and becomes available for analysis. If the expectation is quicker data availability, then edge analysis is where you might turn, where network latency and data transformation will not introduce the expected delays.

Like most modern architecture work, ensuring that the contract between layers and systems is well-defined allows for flexibility as data moves up the stack. In this section, we will briefly describe each layer for clarity and then provide more detail on how different layers can be implemented in theory.

Sensors and device controllers

In this section, we will describe the first two layers. The first layer consists of sensors and actuators for measuring and controlling physical systems. The second layer is made up of device controllers and systems. Initially, our focus is on collecting data at the source. This requires placing sensors within the environment, attaching them to equipment directly, or connecting them to existing available sources. In the next chapter, we will dive into detail on sourcing, configuring, and distributing sensors in the wild, as we explore the concept of environmental monitoring.

We define a sensor as a measuring device that can collect direct analog or digital data from, or about, a system. In reality, it is a transducer, converting energy from one form into another, that measures things such as temperature, or pressure, and provides an analog or digital voltage based on that measurement. Alone, this device does not do much. It often receives some power or a data signal and then returns a resulting voltage or some type of serial code to identify the value it is reading. To make use of that reading, the sensor requires a connection to something that can trigger the reading, read the output (and often power and input values) from the sensor, interpret the value, and send it out to waiting systems. We refer to the thing that reads and transmits the value from the actual sensor as a node, though, in industrial terms, it is often referred to as a **Remote Telemetry Unit (RTU)**:

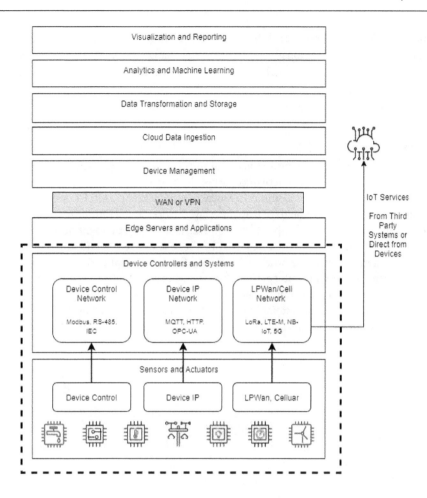

Figure 2.2 – Sensors and device controllers

Figure 2.2 focuses on the initial layers of the architecture. Sensor nodes can often be attached to multiple sensors (or transducers), providing measurement values as mentioned, such as temp, pressure, light, and more. Often, they are closed devices with several sensors included or attached, which communicate to a network or cloud environment as a single device with a variety of measurements within the message. But it is important to keep a logical distinction, to maintain a hierarchy of devices in the network and where devices are attached within the system. These measurements are separate values of different types, and we need to manage and record them so that later, during analysis, similar values can be analyzed or compared as a set. In other words, temperature and humidity are separate measurements and should be treated as such, even if they come from the same device.

Actuators, on the other hand, are control devices that regulate the end devices through the inputs from the layers above. They are responsible for the physical change in the state of the behavior of the end devices. Having the ability to remotely actuate devices and more autonomously helps manage enterprises from a single intelligent command and control center. While doing this in real time, security is a key consideration. We will delve into this later in this book in *Chapter 9*.

More or fewer layers

During this book, there will be many architectural and convention-type questions to consider. For example, are industrial control networks part of the sensors layer or the controllers layer? Do we add a layer in between to accommodate local controllers such as PLCs or closed system networks such as SCADA? Or should we map these layers closer to the Purdue model to provide better consistency with reference architectures? This is highly dependent on your existing environment and architectural standards. It may be the case where you add multiple layers to accommodate different control networks within your factory. In many cases, your industrial network is separate from your enterprise network and this will affect your design. In this chapter, we have presented the simplest case that can be modified or built upon, to showcase interacting with your industrial systems and to facilitate data flow and analysis. Consider this the most basic case that will fit most environments, but expanding this architecture is, in most cases, almost certainly required.

Automation patterns

We have broken the first two layers into distinct types of systems defined in the following list. This will help us define an initial set of patterns, based on the type of sensor or network, and the way that data is being accessed from the factory floor. For large organizations, the goal is to define patterns and approaches for accessing data across a large factory or many factories. This can get complex, but if you can standardize the hardware, software, and data acquisition based on the type of device, machine, or assembly line, you have won a large part of the battle.

In the long term, this can simplify the deployment, monitoring, and management of your instrumentation infrastructure. For those not as familiar with **Operational Technology** (**OT**) and what you may be facing, these major types of environments consist of the following:

- **Industrial control systems**: Most commonly, we think of SCADA as being in this group. **SCADA** is an acronym for **Supervisory Control and Data Acquisition Systems**. It is commonly used in conjunction with additional control equipment such as **Programmable Logic Controllers** (**PLC**) and RTU. Together, this equipment with a fieldbus industrial network system using a defined protocol such as Modbus, BACnet, or Profibus is used to control large or distributed industrial systems in real time.

    These systems are often closed-loop systems, especially legacy systems that have been around for a few decades or more. Getting data out of these systems can require translating the data from one format into another to allow data to be sent to external systems such as the cloud.

Technically, this is much easier today with the prolific amount of hardware and software that caters to this type of instrumentation. There may be internal challenges within the production team as to what is instrumented and how with the number one priority being to avoid any disruption to ongoing production.

- **Device IP networks**: Device IP networks are more modern in their technology approach, interfacing with IT networks using common protocols that interface more easily with standard IP or TCP applications. More commonly, we think of devices or systems in this category that already speak in protocols that we can readily consume. MQTT comes to mind in this category. It is a simple, lightweight protocol that already relies on TCP to send messages. The message header is composed of simple strings, but in many cases, the data packet can be encrypted or encoded for security or size constraints.

  Multiple vendors sit in this area, either interfacing with, or driving industrial control systems, or sitting beside those systems to provide instrumentation. When diving into integration, it becomes a necessary exercise to evaluate existing systems and determine if integration points are currently available. In many cases, vendors are working to provide a complete solution for digitizing your environment. If this is the case and you are open to it, then the problem is solved. But the reality is that no one does everything that is needed and integrating from a diverse set of components is the brutal reality of many operations teams worldwide.

- **IoT networks**: IoT networks such as **low-power wide-area networks** (**LPWANs**) can be a broad category and are designed to incorporate sensors and nodes that do not connect to traditional wired networks within an industrial setting. There are several recent new wireless protocols and systems available that can allow instrumentation without ever touching the internal working of the equipment. Some of these have been around for several years and have been vetted in hard environments. These devices are generally low-powered and can vary in terms of distance and the amount of data they can send from each node.

  Long Range (**LoRa**) or **LoRaWan** has become a popular option, where the amount of power used or the size of the device must be considered and where range (up to several kilometers) is needed. An alternative such as 4G or 5G can be a good option if a public network is required, such as in an outdoor or urban area, and you can't put gateways in place. We will speak much more about some of these options in the next couple of chapters.

With all these options, things can get very diverse very quickly. That is kind of the point. How do we approach legacy and various environments in a structured manner? If you are building a new factory today, much of this instrumentation can already be built in, along with integration into manufacturing execution systems, which helps drive operational control and efficiency. But there are thousands (millions?) of existing industrial farms, factories, and plants that can benefit from further digitization. Even in the former case, with modern integrated machinery and data acquisition, we still need someplace to transform, store, and analyze this data. Getting data from existing systems, while sometimes a challenge, is possible. But the question of *now what?* takes over, and we need to look at what to do with the collected data.

Edge servers and applications

In contrast to the previous sensor layer, which in many cases can be narrowly defined, the controller layer is much broader encompassing everything from simply defined ladder logic within PLCs to full-blown edge servers that analyze and pre-process in-flight data streams in real time. We mentioned this briefly in the *More or fewer layers* call-out – additional layers may be required to accommodate different manufacturing or controller networks and systems based on your environment:

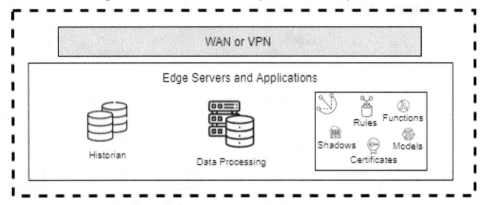

Figure 2.3 – Edge servers and the applications layer

Simply put, we can view this layer, outlined in *Figure 2.3*, as where the data is extracted from the environment and prepared for forwarding or processing at the next layer or in the cloud. This can be done by interacting with an existing set of systems such as a Modbus interface, by interacting with a local historian or SCADA interface, or even by receiving sensor data directly through a local wired or wireless network. Edge devices such as controllers and servers can serve different purposes, such as the following:

- Assist with getting data from proprietary or non-IT protocols into a more IT standard format, such as MQTT.

- Reduce latency issues in getting on-premise data to the cloud, due to the potentially high volume of data that can be generated. Data can be stored for later transfer, preprocessed, aggregated, or filtered as necessary.

- Perform on-site (edge) analytics and control based on current conditions in a more real-time manner.

Traditional closed controllers such as PLCs or PACs may also exist in this layer, or within the environment, with adaptors receiving data from these devices for forwarding and analysis by operations. Not all devices in this layer need to be smart devices. Some are used to simply interpret, translate, or forward data to the cloud. This is more often than not the actual case when starting out or with lower priority systems. We simply want to visualize operations in a more meaningful way to monitor consistency, throughput, or equipment-level metrics.

WAN or VPN to the cloud

While not precisely a layer, WAN is defined within the architecture; we must take a moment to discuss the logical and physical divide between on-premise data collection and analysis within the cloud. The divide between the factory and the corporate network (or the cloud) can be simple or complex, depending on how enterprise IT has structured the network interfaces and security implementation. Often, there is a defined path for data flow from inside to outside the factory walls, and it is vital to define this transport securely. The goal is to protect both networks from intrusion and ensure data integrity.

Even in cases where you are using a third-party transport such as LoRa or 4/5G, you can control the egress of data where it flows within your network. In some cases, where more public transport is used, such as LTE, this may not be possible, so physical security is essential for your endpoints, and you have to rely on the security of the third-party network to protect your data.

Device management

We have broken device management into a separate layer, as shown in *Figure 2.4*; however, it may not necessarily need to be based on your environment. Device management focuses on onboarding, managing, and securing your devices in a simplified manner. With hundreds or thousands of devices, managing and organizing your devices can be a handful. This is especially true if devices change frequently. Certificate management can also be handled in this layer, especially with intelligent edge systems where you want to ensure those devices are secure before communicating to your platform:

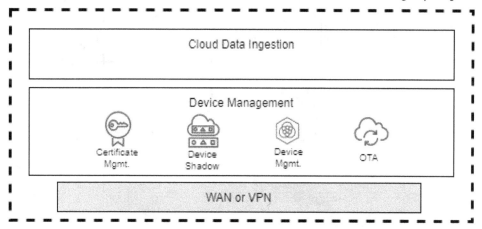

Figure 2.4 – Device management layer

Device shadows, also called digital twins in the industry, are another approach that you can use to track and manage your devices. A device shadow is a digital representation of your device that maintains its current state and configuration. This allows you to interact with the shadow instead of the device directly and apply changes to see how the device reacts or to push those changes to the device.

A device shadow will also have the last known configuration and state of the device, even if the device is offline. Users can see what that last configuration was before it went offline or disconnected from the system.

Cloud data ingestion

This is where data collection starts for the cloud. Ingesting and interpreting data in a variety of formats can be a challenge. But if we break it down and identify patterns, it becomes manageable and maybe even fun. Our focus is on collecting data from factories, fields, and equipment that are used in manufacturing and industry. The type of measurements collected can be wide-ranging in their type, volume, and importance of the data to operational and business processes.

Ingestion patterns will probably be specific to your organization and your data but not uncommon in the industry. AWS provides several services for you to ingest and transform data, and depending on the format, protocol, volume, and velocity of your data, this will drive some of your decision-making in this area. There is rarely a *one-size-fits-all* approach, especially for industrial solutions with lots of legacy solutions, and often, one approach is as good as another, so long as forward progress is being made. Often, it can depend on your experience, skills, and preference in these technical areas:

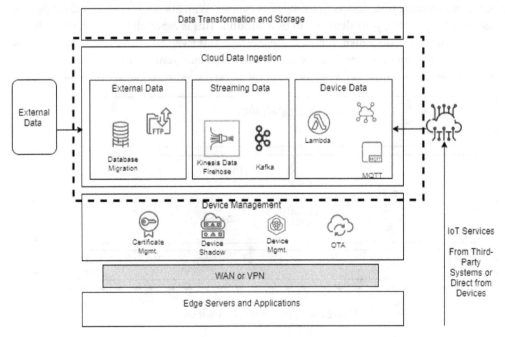

Figure 2.5 – Data ingestion

Figure 2.5 outlines some of the data ingestion approaches available. There is another gray area between data ingestion, and data transformation and storage. Do you store raw data directly before it is transformed in any way? This can provide a sense of maintaining data consistency and accuracy, to be able to trace back to the original data when something seems wrong, or the information is challenged. There are many thoughts concerning populating raw data storage:

- Consider how much transformation is made at the edge before data and measurement values even make it to the cloud. Activity at the edge is often used to reduce bandwidth or increase response time with high-volume data streams. The integrity and accuracy of data might be challenged at lower levels in the architecture.

- Consider if data should be lightly transformed before being stored as raw data. Some data comes in as binary or serial data strings, encoded in some way to reduce the size of data in transit. Should this be stored as-is or converted into readable text and values before being stored?

- Finally, consider the cost and effort of storing raw, untransformed data in the long term. The benefit may far outweigh the cost if data is called into question and can be traced back to the source.

Alternatively, there may be legal requirements that require this information to be stored, raw and unchanged. This could be mandated and hence is on the table for your strategy. There may, however, be different requirements for different data streams, but the best is to define your raw data storage requirements and standards and then keep them consistent.

Data storage and transformation

Storage and retrieval are critical areas to consider, especially with large amounts of data. The options are many, and this needs to be thought through considering your current and future (unknown) needs. Some options include:

- Data lake

- Data warehouse

- Lake/warehouse hybrid

Let's look at each of these in a bit more detail.

Data lake structure

Using a data lake as a data repository provides good flexibility and the ability to store all current and future data in a way that is cost-effective, easily manageable, and useful to transform for future analysis. You can think of a data lake more as an object or flat file storage approach (think a Windows folder structure) versus a data warehouse, which is more relational. The concept of a data warehouse has been around for decades and is still a very common and useful way to store and analyze relational data. The main drawback is that data has to be formatted and structured in a way that it fits into the data warehouse successfully, whereas a data lake simply stores any type of file in any type of format.

In AWS, the most common data lake approach is to use **Simple Storage Service (S3)** and store your data as flat files within a *very* well-defined folder structure or hierarchy. In addition, you can and should use multiple areas within your data lake to store and transform data in different stages as it is processed. You can think of these as different lakes with streams attaching them, with the water getting cleaner as it flows from the first lake to the last. Many layers of processed data can exist within your lake, but we will narrow it down to a few of what we consider the key layers:

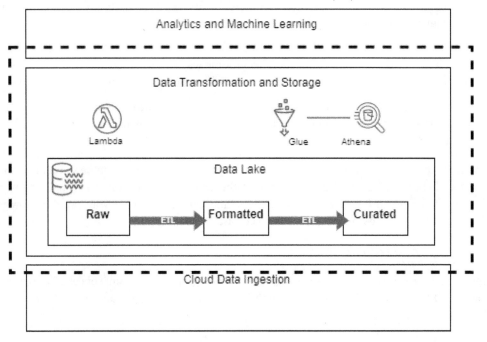

Figure 2.6 – Data transformation and storage

Figure 2.6 outlines one approach for designing your data lake environment and storing your data. Let's break down this approach in more detail:

- **Raw data layer:** The raw data layer is where all data is initially captured or ingested. This data is generally not ready for consumption or use for analysis. The natural benefit of this layer is that data can be quickly ingested and stored in the raw layer. This can make it much quicker and easier to design the intake of the data stream since little transformation is done at this point.

  There are several common rules or standards set against the raw data layer. Following these conventions is optional, but for some industries that need strong traceability, and to ensure the integrity of data, these basic rules are useful:

  A. Data is stored in its original or raw format. Some message analysis may need to be done to determine where in the lake the data should be written. However, the raw original message is what is stored with all metadata attached.

B.   Data is immutable or cannot be changed once it is stored. Additionally, this lake should be secure. It does not seem likely that access to this data is required, so lock it down and protect the integrity of the store.

A potential design point arises around the folder structure of the storage, or schema if you will. This needs to be well-defined, ensuring that data can be found and managed easily as data grows and more sources contribute to the lake. The format of the structure can be flexible and should be defined based on your needs. Time series data can take a familiar structure to allow you to find specific data by type or source; for example, Source → Component → Type → Year → Month → Day → Hour → <Minute>:<Second>. However, your needs may vary, depending on the type of data and your industry.

Defining a schema

Structuring your data is a common struggle with any type of storage approach, whether it be a data lake, a relational database, or even a time series database. There are many examples on the web, but different approaches are commonly based on your needs. Coming up with the right structure or schema for your data is one of the first tasks that need to be accomplished, and often, that is when you have the least amount of information about how data will be acquired or used. Our advice is to move slowly and carefully. More than likely, you will not make a wrong or disastrous decision, but it is easy to second guess your decision at a later point, and end up spending time and effort to remove technical debt with a restructuring effort.

- **Formatted data layer**: Initial transformation and formatting should occur when moving data from the raw to the formatted layer. As soon as data is deposited in the raw layer, this transformation process can begin. This means the data can start to have meaning and structure as it is consumed in this layer.

In the raw layer, data is still in its native format as much as possible, but as we format, transform, and standardize the data, this goes away and a 16-byte hexadecimal string is transformed into a set of real values such as temperature and humidity. The schema or folder structure is similar but more granular as we store individual measurements with DateTime stamps for easier curation within the next layer.

The ETL from raw to formatted is probably not complex but could be widely varied if you are receiving data from a variety of systems and solutions. Each type or source may have a different encoding scheme for measurement data to ensure the smallest bandwidth use possible. With good encoding, multiple measurements can be stored in a very small string that's passed over a serial or wireless communication path. However, decoding those strings will be specific to each type of data packet.

In addition, data can be restructured at this layer, moving from raw strings to a columnar data format if necessary. Some of the data within the raw data layer may already be in a useful format, but truly raw data in its original format can be transformed and rolled up into more useful file formats such as CSV, Parquet, Apache ORC, or Avro. Each format has advantages and disadvantages based on the amount of data, and how you might use it within your analysis. We can also strip away some of the metadata that may come with the message if it is not necessary for subsequent analysis.

Access to data in this layer may also be restricted based on your data governance policies and analytics needs. The value of this layer is to quickly convert your data into a more readable and structured format for easier consumption.

- **Curated data layer**: Finally, we come to the curated layer. This is where we start to structure data in a useful way for analytics and the business. Formatted data can be restructured in this layer and combined, aggregated, or transformed based on need – for example, when additional values need to be calculated based on existing raw measurements. An excellent example for the agriculture industry may be to calculate **evapotranspiration (ETO)** based on weather data that has been received over the last 24 hours. This is a highly complex calculated value derived from raw input that would be beneficial for data analysis.

  The structure of this layer can be quite complex as the curated layer may also combine data from other sources, such as financial or business systems. Data from this layer can be reformatted and fed to a data warehouse or analytical methods. We can expect data to be denormalized in this area as the needs of the business become better known and can be moved into more traditional database systems, data warehouses, business analytic systems, or machine learning models.

  Access to this layer provides the most freedom within the organization but should still be restricted based on need. Most of the analytics will be performed by moving data out of this layer into external systems or databases, which provides an additional layer of abstraction into the data lake itself.

Three-layer approach – more or less?

In this section, we outlined three main layers within the data lake architecture; however, this is just one example of how you might create your data lake structure. Some people prefer to name their lakes Bronze, Silver, and Gold. It is also perfectly OK to have more or fewer layers, depending on any number of factors, such as the size and type of your data, or any legal requirements you may need to consider. We say this again and again within this book, *one size does not fit all,* and, as architects, our job is to research and understand the best-of-breed approach and reference architecture and then apply them to our organizational goals.

Metadata management

It is probably the case that you will need to store some information about your data, such as the source(s) and destination within your data lake, refresh intervals, type, format, life cycle, transformation information, and security considerations. Additionally, you will need information about the sensors, types, and locations. For example, you may have some data stored under a specific component ID; however, there is no information in the data stream about where that component is located, how old it is, or which factory it belongs to. A set of metadata can be used as a lookup when necessary to get all this additional information about your system.

A database is the most likely place to store this information and can be easily queried to figure out where to find data that is relevant to your needs. There is a more recent approach called a data lakehouse, which combines the stability and flexibility of a data lake with the query ease of a data warehouse. This approach can have many benefits and should be considered if you are getting started on your IoT initiative. However, you can custom-design your set of tables using AWS Relational Data Service. Depending on the scope of your initiative, it may be helpful to build a frontend to your metasystem to enable the easier deployment, configuration, and management of your metadata.

Analytics and machine learning

Moving one step higher in the architecture, we can get to more of the fun stuff. Much of the work in the previous layers is plumbing and wiring, which is extremely important and can also be fun (for nerds like us), and it is necessary to get it as correct as possible, given what we know (or don't know) this early in the process.

However, once we have started collecting data from our systems – and in some cases, it may be an enormous amount of data – we want to quickly move to the next step and try to understand the data better, and hopefully use it to improve our operations and production and quickly show proof of value. In *Chapter 1, Welcome to the IoT Revolution*, we talked a lot about the business side of using data and making sure we are collecting the correct data for the right reason. Otherwise, we may be left with high data storage and processing efforts with minimal business benefit:

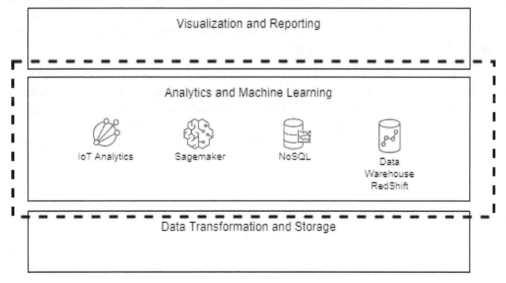

Figure 2.7 – Analytics and machine learning

There are two main areas of focus in this layer, as shown in *Figure 2.7*. They are broken down because they can have distinct goals.

Machine learning

Machine learning is an exciting and relatively new topic that many organizations view as one of the top goals for adopting an Industry 4.0 or data-driven strategy. In an industrial setting, there are a couple of goals we hope for when implementing machine learning:

- **Machine sensor data analysis**: The goal here is to present a data stream and have a machine learning model to baseline that stream and look for variances, hopefully alerting us to potential problems that need to be examined before they become significant issues. Ideally, this analysis could incorporate multiple streams of data (or numerous measurements) to determine how the system components are working in concert together and if there is a correlating problem in the equipment.

- **Video or image analysis**: Image analysis is similar, except that it reviews a video feed or static images and determines if there are anomalies in the product presented within the image. Things such as colors, placement, or surface issues are detected, and a determination can be made if there is a production line problem that needs to be addressed.

There are other types of models of course, and we will touch upon them in *Chapter 12, Advanced Analytics and Machine Learning*. For example, consider a demand water balance model for a large utility that processes several factors, including weather, demand history, supply, costs, and more, to predict demand at the consumer level for a future period, thus allowing the utility to optimize its water sourcing strategy and minimize cost. This type of analysis is much more advanced than what we can describe here, but requires both sensor data from flow meters and demand systems, along with many external variables to augment the analysis effort.

In addition, image analysis is not something that we will cover in this book. It is a specialized topic that is well covered in online sources and other available books. Data streams, however, are something that we need to understand better, such as what is available and how we might approach achieving our goals in this area. It is relatively straightforward to train a model so that it detects objects or anomalies in an image and can provide a tremendous benefit in evaluating output quality early in manufacturing processes.

Business analytics

There is a difference between business analytics, which is future-focused, and business intelligence, which looks more explicitly into the past. It could be argued that this analysis should be moved into a higher layer since it is an activity that is performed by an end user and, more specifically, a business analyst. The analytics goal is to use data to generate insights about how things are running, how they will run, or where problems can occur.

Early efforts to find efficiency

One of the earlier IoT projects for one of the authors was with a large golf club in the Southwest desert. This particular club was world-class with multiple courses, several of them PGA-recognized. The irrigation system was quite complex and water moved around and between the different courses into ponds and streams, from where it could be used for irrigation and keeping the courses and the landscape in top shape. Once the instrumentation was in place, which included monitoring everything from water flow rates, pump data, lake levels, and weather data, to the soil moisture on the courses and greens, there was a great system for viewing real-time system data. Our ultimate goal was to find efficiency, reduce waste and cost, and look for anomalies in the system as they occurred.

Anyway, at one point, during those initial stages, we exported data into Excel and built graphs, trying to understand system relationships. We then discussed our analysis with the customer to see if this was interesting to them – things such as water flow versus power usage to determine pump efficiency or possible deterioration. The greatest challenge was trying to estimate green hardness (this determines how fast the ball will roll) and how the weather, the amount of irrigation from the night before, and more will affect it. This was seriously a hard thing to determine since there were so many variables involved. And those algorithms are still being tweaked today.

Machine learning often starts with analytics and learning to understand the data and relationship to know what is normal, what might be improved, and what the data looks like before, during, and after anomalies occur.

We could switch or combine analytics initiatives within the visualization and reporting layer. At the very least, let analytics and visualization work hand in hand to provide better and more complete information. Analytics provides the analyst or business users with the tools required to better understand what is happening from a production level. Information can be combined with other plants and corporate systems to determine where improvements or changes can be made. How many widgets did we build and ship last month, and what is our prediction based on how the systems are running? This is a pretty basic question that can be determined relatively easily.

But beyond that are some of the more enlightening questions concerning the underlying systems that manufacture those widgets. Are they slowing down or heating up? Do we need to request early maintenance to get them back up to spec, or can we push them a little longer? Instrumentation and baselining your equipment will help you answer those questions. And providing clean and valid data with the right tools allows you to ask the right questions.

Edge analytics

Although edge computing and analysis officially reside in a completely different layer, it deserves a quick review. The benefits here are well known:

- **Reacting in real time**: When things are happening fast, we cannot wait for latency issues that may occur with sending data to the cloud. In some manufacturing processes, milliseconds count. If something is wrong or gets missed, it could mean thousands or millions of dollars in equipment repairs, or manufactured goods that need to be discarded due to quality issues.

- **Bandwidth slowdowns**: Again, this can be a function of high speed or specifically the high frequency of measurement data. Consider a sensor that measures pressure a few hundred times a second. Now, multiply that across one factory. This can be a lot of data you are trying to send to the cloud. Pre-processing or storing that data locally makes sense based on your use cases for review and analysis. It might be stored and moved via another route to avoid bandwidth for higher-priority data.

I mention these ideas because some analysis is essential and needs to be done quickly and efficiently, while some can be done at a more leisurely pace. We will consider this more when we focus on edge processing and analytics in future chapters in more detail.

Visualization and reporting

In many cases, users just want to see data, especially early in the program, and usually in a graph format with the ability to view different time scales or compare different measurements or areas with one another. This could be raw data formatted just enough to make it graph-ready. The visualization and reporting layer is depicted in *Figure 2.8*.

This concept of viewing unfiltered data from your systems is used during the initial stages to help you better understand the environment. What does *normal* look like? I find this question fascinating. Operators often know what is normal, but not always the specifics. They can hear, smell, or feel if something is wrong, but they cannot accurately define or measure it. Once a new insight into an operating environment is gained, once that environment or equipment is instrumented in new ways, knowing what is normal becomes much easier and specific. Once that insight is learned, over time, an operator or analyst can better define and detect anomalies, and this can be built into machine learning models:

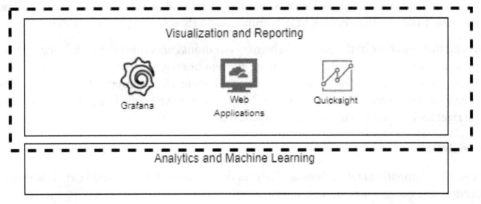

Figure 2.8 – Visualization and reporting

Even with basic data analysis, much can be done. Take a simple pump as an example. Basic questions can be answered, such as energy versus flow rate. Layer in energy costs, and now you can determine how much is saved if you reduce or increase the flow rate. From a reporting and business intelligence point of view, there are two primary options:

- **Report**: A report provides a view of data with additional tools that enable the user to drill into the data or view it in different ways – for example, to view the data in a more granular way, such as at the second or sub-second timing period. Analysts can use reporting tools to answer questions about machine operations, look for abstract anomalies, or decide where plant improvements can be achieved.

- **Dashboard**: A dashboard is a little more fixed in the sense that it shows the status of the systems at the moment in time. Often, a dashboard can be set up in operations, where potential problems can be detected to determine which locations are active or make other choices about the equipment in real time.

This is not new to those familiar with BI and reporting tools, but it helps to know your target to plan your way forward and share information and progress. This will allow stakeholders and those dependent on the data to understand and help drive the project forward in a meaningful way.

Custom applications versus OOB

There are many **out-of-the-box** (**OOB**) products that include visualization and reporting capability as part of their solution. It can be tempting to start with an existing solution you already have available, such as part of your ERP, asset management, EA, or perhaps another product. There are several things to consider with this initial approach.

Many of these products have great reporting or analysis capabilities, but mostly within their own data domain. Incorporating data from other sources can be a challenge. For example, streaming data can come from sensors or data acquisition systems, or direct from the historian. This data can, in some cases, be combined with the order, cost, and usage data to help understand the potential cost/benefit when anomalies potentially occur. Additionally, maintenance history and effort may be layered on top to help you understand when and what maintenance last occurred, or if it is scheduled for the near future.

All of this, when combined, may be difficult for any OOB solution you are considering. However, it may also be cost-prohibitive to build such a solution from scratch. As an architect, these are the choices we need to analyze and determine the best fit solution for our organization today and into the future.

How much custom development is required?

In previous projects, the authors have built some pretty complex and custom business intelligence or analysis products from scratch. This was driven by a couple of key goals:

- First, IoT was still pretty new, and performing analysis on multiple but related time series streams of data was something new to most reporting tooling at the time.

- Second, the goal was to provide an easy-to-use interface that allowed users to view different data streams and different time zones across multiple plants or regions. We did not expect our users to be sophisticated, or to have to learn a complex BI tool.

- Third, we were building commercial products that would be offered to customers as a product or service. With custom development, we didn't have to worry about license costs or embedding problems.

Today, the world is a little different. Many industrial control systems and third-party products provide much of this type of analysis capability. In addition, there may be different types of users within your organizations, analysts who can learn and work with complex tools to better understand patterns in the environment and business or operational users who may be more interested in dashboards and graphs showing what happened last week, and what will happen next week given the current plan.

Bringing it all together

That's it so far – for the layers at least. This reference approach should give you a feel for what you need to consider as your starting point. It is far from the final architecture, but if you don't have any idea where to start, it hopefully provides something to think about.

We have tried to cram a lot of information into this chapter to help you understand this reference architecture and some of the reasons why it was put together in this manner. Everyone has to start somewhere, and without knowing much yet, about how we are going to solve some of our digitization efforts, this is a good starting point. We can now focus on implementing parts of this architecture and use it to test out our solutions and standardize patterns.

Figure 2.9 outlines the complete strawman architecture. It is a lot to consider at once, but we have tried to keep it high-level and consumable for most people. We are only scratching the surface:

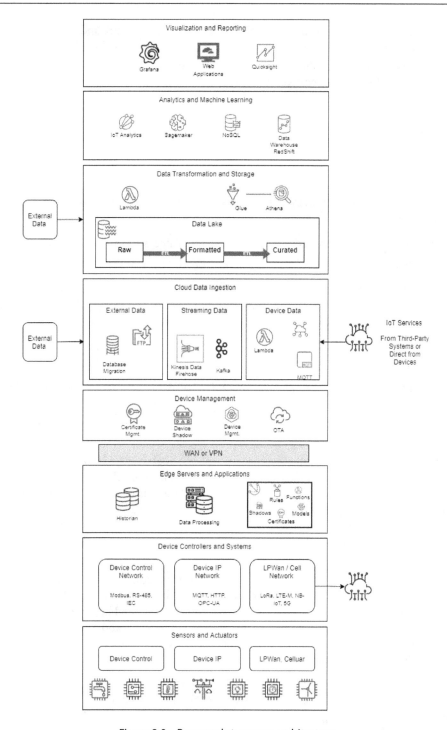

Figure 2.9 – Proposed strawman architecture

We will switch gears in the next section and talk a little about architecture and development best practices. This is an architecture book after all! But really, we want to talk about a few topics that will help as you continue down your path. We all get busy, but at this stage in our project, it is too easy to start making big mistakes that can cost you down the road.

Defining standards

As architects, we have multiple goals:

- Defining the architecture and product selection is a function of serving the business and trying to serve and enhance the requirements and goals of what the organization and business are trying to achieve

- Defining an architecture that is functional, robust, scalable, secure, and yet, as cost-effective as possible

Another piece of the puzzle is setting the standards and conventions for the development and implementation teams to follow within the organization. It is equally as important that the organization have these standards that can be applied before any real effort begins.

As an enterprise architect, you may not be doing code reviews with the team; however, you should highly consider setting some guidelines for application architects and team leads to follow, as well as ensuring the project is scoped to allow time for more rigorous software engineering practices to be followed.

> **The role of standards**
>
> Several years ago (OK, many), one of the authors published a book called *Application Architecture for WebSphere*, where a whole chapter was devoted to setting and following standards and conventions within a project and team. At that time, the cloud was relatively new, so everyone was building and deploying applications in-house or within private data centers. However, many, if not most, of the guidelines that were described back then still hold today in some fashion, requiring some modification in the application to provide updates in technology.
>
> The main point here is that *standards* are really important, especially for large teams or long-running projects. We all know the stories about code that shouldn't be touched, and we also know just as well that a little bit of foresight could have helped to avoid these issues. As an architect, it is our job to reduce potential issues, ensure code quality, and help our project nicely live into the future. Not indefinitely, mind you (that is another story about over-architecting because of some abstract potential use cases), but into the foreseeable future as many things continue to live well beyond their natural time.

This means ensuring that standards are followed, code and infrastructure are well reviewed and documented, and the goal of the project meets existing requirements. Our job is to pass on to the next development, or architect, a project that you can be proud of. To start, let's define some simple concepts:

- A **standard** is a strict rule outlining the way something should be done.
- A **convention** is more of a suggestion. It is a guideline for how to do something, but it doesn't need to be followed exactly.

Standards are probably one of the most important aspects of building applications and environments yet are just as much overlooked as an opportunity to achieve your goals. There are guidelines aplenty on the internet and putting together a reasonable set of rules should not be difficult for experienced architects who understand the chosen set of technologies.

In the software industry, we have learned a lot over the years, along with knowing it is not always possible to follow strict standards and still meet the goals that the business or project has dictated. Managers want it faster, cheaper, and better, and the marketing machines of many vendors have convinced them this is possible. If you are following agile – and let's face it, who isn't? – things can change quickly, and with short sprint cycles, strict code reviews are not always possible.

> **A comment about methodology**
>
> The statement about agile is a little skewed since we all know there is a right way and an incorrect way to adopt any practice or method. We wouldn't want agile or methodology pundits to think we don't know the difference. We do, and we cry (and possibly something bad happens to kittens) every time it gets twisted into some cruel form of punishment to get more done, faster, from a development team. It's wrong, but it's often the reality of development projects today.

Basic project review recommendations

At a minimum, there are some pretty easy things you can do to assist your survival on long projects. I'll mention a few here to start you on your way. These are things you could do once a quarter or maybe twice a year, helping to keep the project aligned if you can't or won't provide heavier oversight in the development process:

1. Take a look from time to time at your source code management system:

    - Do the project and repository names make sense?

    - Are the names consistent?

    - Is there a brief one- or two-line description about that set of code explaining its use?

    - Is the main description page current and does it do a good job describing the module and its usage?

    It is no fun when new team members come online and can't make heads or tails out of the 300 repositories on your project (yes, we have been there).

2.   Review your architecture documentation:

- Are your overall diagram(s) accurate and up to date?

- Are there major flow changes that need to be updated?

- Can you tell from the documentation how things are structured, and what major components are in use in the system?

Things happen fast in this space, and the architecture can change considerably over time. You need an accurate set of diagrams to review with your team and share with others who are asking about your project. Updating constantly is a big ask, but twice a year should be achievable, depending on the amount of churn in your environment.

3.   Meet regularly with the entire team to discuss concerns about the quality of the product. Allocate a short amount of time, perhaps one day every month or quarter for a round table discussion or maybe breakout sessions to fix minor issues. This team may include only the development members, but this can be expanded to include architecture, management, and infrastructure members to share issues or improve working processes going forward:

- Are there things that we can do better?

- What is out of alignment that we can fix today?

Make this non-combative and focus on what the team can do to improve. Often this can be done during a sprint review regularly, but to work, the feedback has to be incorporated into the sprints ahead, which may affect velocity.

Additionally, you can use tooling to ensure that standards are followed, or conventions are used. Static code analysis tools can be very helpful in ensuring some type of code quality. If you are not a code monkey, please remember that 100% adherence to static analysis is not always possible, or if so, then it comes at a price. So, use these tools judiciously to ensure your effort is cost-effective, as well as the highest quality.

Setting expectations

These guidelines are not magic, and management needs to be aware and involved in every aspect of the constant battle between turning out a quality product and doing it quickly and cheaply. This is not always possible since, in large development organizations, there is a separation between development, architecture, and product management. This is by design so that one team cannot make decisions at the expense of another. In companies where development is a by-product of doing business, this is usually never the case. Management holds all the cards, both good and bad, depending on the project's outcome.

There is value in building a good relationship and trust with management. And there are sometimes tradeoffs in the amount of oversight (and potential quality) of something versus getting it done quickly. We must find that balance in our projects to ensure that some quality is allowed to show through while we still achieve our delivery goals.

The information and guidelines presented so far are not secrets – they are pretty basic tasks. We outlined them here to help you get a grip on things as you move ahead faster than you will likely imagine. If you build some of this into your routine, it will help you multiple-fold years from now. This doesn't excuse you from setting and following more rigid standards and conventions. Still, if time and resources do not allow, and the business pressure is too high, you can cover these with at least a minimal set of activities.

Summary

This was a lot to cover in one chapter, and we are just getting started. If you are new to IoT and architecture, we tried to keep it meaningful and understandable. If you are an old hat at this stuff, then hopefully, this was a decent refresher or allowed you to consider some alternatives to your current thinking.

At the end of the day, however, we have a good starting point, and we can build upon this as we walk through various examples. This architecture will evolve as we put some of the pieces together, either swapping in new services or connecting services in different ways to accomplish our goals. Our advice is to have fun with it for now as you connect the dots and consider alternatives based on your environment and your needs.

In the next chapter, we will start to look at collecting data from basic IoT devices in various ways. Our initial focus will be on environmental monitoring, which means deploying sensors directly into your environment to measure more than just machines, whether that be a factory, an oil and gas pipeline, or a farm or forest in a remote location. Remember, industry can be anywhere, and many of the lessons learned in one location can be used in many other situations.

3

In-Situ Environmental Monitoring

When we think of industry, we often envision factories or large processing plants; however, industry is more than just manufacturing. We can expand our definition to consume other branches of the economy and the production of materials or goods. In addition, service classes of industry can also be considered. In most cases, digitizing these environments and transitioning to a data-driven approach can improve operations and efficiency for many everyday processes and activities. And finally, there are just remote and hard-to-reach places where instrumentation and monitoring are useful (think cold chain management).

Before we delve into manufacturing and machine monitoring, we want to look at monitoring solutions more holistically. In this chapter, we will explore several common industries and use cases for environmental data collection, focusing on the approach and the ultimate value we can derive from our data.

In this chapter, we are going to cover the following main topics:

- What do we mean by environmental monitoring?
- Integration patterns and protocols
- Getting data from the field
- Data drives everything – a common use case
- Wired versus wireless
- Gateways versus intelligent edge machines

Manufacturing in the bigger context

In the previous chapter, we delved into the realm of architecture and system design and the constructs, layers, and importance of a flexible design approach. Armed with this knowledge and using architectural thinking for a contextualized problem, we could arrive at a solution that fits a particular situation unique to your manufacturing industry or facility. *Figure 3.1* outlines some of the primary functions within the manufacturing process, with production being the core function of any given manufacturing facility:

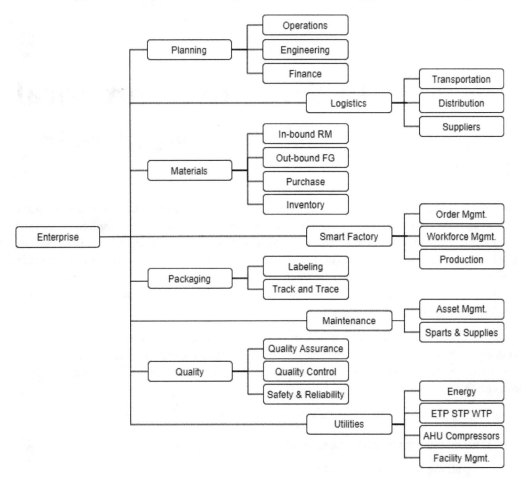

Figure 3.1 – Smart manufacturing value chain

But there are multitudes of supplementary and complementary functions that go seamlessly to facilitate on-time production, near-zero downtime, and maintain the high-quality output of manufactured goods:

- **Planning**: Planning involves product engineering, design, operations, and finances.

- **Logistics**: We have the logistics division accounting for the timely inbound transportation of raw materials across suppliers, vendors, and geographies.

- **Packaging**: There is also an outbound supply chain for shipping the processed or finished goods to the end user.

- **Materials**: Then comes the material function involved in storing, indexing, and managing the inventory of raw materials, as well as finished and semi-finished goods. After this comes the core function of manufacturing, which involves machines, robots, and the workforce responsible for actual value creation. However, the buck doesn't stop here.

- **Quality**: We have other essential functions such as quality assurance, packaging, maintenance, and facility management.

Each of these has many subsystems under them, as described in *Figure 3.1*. All of these co-dependent systems need to function coherently to create value with increased efficiency for the business. All these functions have a fundamental underpinning that drives us toward the core concept in this chapter. Physical environment and parameter sensing are imperative to many parts and sub-functions. Quantifying the various **Key Performance Indicators** (**KPIs**) and exchanging this information with adjoining functions is essential for coherent operations. Sensors are the magic blocks that aid in gathering this information.

As seen in the previous chapter, sensors can be defined as devices or instruments that aid in measuring changes in physical parameters in real-time. They convert physical parameter values into equivalent electrical or digital values to enable integration with a computing device. Sensors are interfaces between the material and the digital world. They quantify physical and operational parameters or characteristic values to human/machine-readable data.

Sensors can sense conditions in or around the devices or objects where they are deployed. The following sections provide a deep dive into the five W's and one H (5W1H) of sensors: what, why, when, where, who, and how.

What do we mean by environmental monitoring?

Sensors play a critical role in environmental monitoring. They help to measure a quantity and provide a signal, typically electrical, as an output. A sensor can be considered an intersecting coordinate of a two-dimensional graph with two varying parameters, such as temperature and resistance or distance and time. Based on the output, sensors can be further classified as analog (continuous), digital (discrete), or binary (zero or one). With the advancements in silicon nanotechnology and digital electronics, sensors have rapidly evolved, reducing in size, improving in accuracy, methods of sensing and speed, and so on. Integrating the compute, storage, and transmission modules within the sensor unit has blurred the boundaries of the actual capability of the components.

Figure 3.2 generalizes the types of sensors pertinent to the industry with the property as a basis of classification. Each fulfills a specific role in quantifying a manufacturing ecosystem/process and creates data consumed by decision control systems along a Purdue-type model. While this is not an exhaustive list, it provides a general classification for choosing and selecting sensors based on application requirements. Biometric sensors such as iris scanners and face, fingerprint, and speech recognition systems have particular usage in the workforce and security-related applications. Most

OEMs also bundle multiple sensor units to arrive at a sensor complex. For instance, to ascertain asset condition, noise, vibration, and harshness sensors are combined as a single module.

In the previous chapter, we discussed monitoring systems that can reside within the machine network. We also identified the difference between the sensor itself, which can take a reading within the environment, and the rest of the system, which captures that reading and transmits it along to the following system.

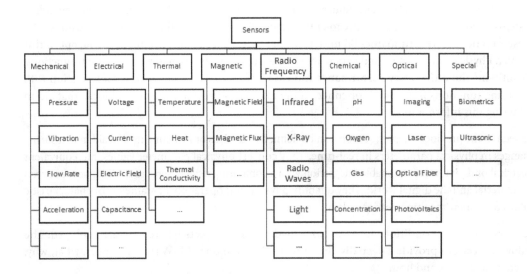

Figure 3.2 – Sensor classification based on properties

Environmental monitoring is the ability to acquire physical parameters in-situ or at-site. In-situ means explicitly putting a sensor in place. Perhaps there is a need to measure the liquid depth within a tank. A pressure, ultrasonic, or float sensor can be used and put in place directly on or in the tank to perform this measurement. Some sensors are already embedded in machines that transmit data to connected controllers/**Programmable Logic Controllers** (**PLCs**). Retrofitting additional sensors is also possible in a brownfield scenario to gather more data to solve the chosen problem.

We often think of in-situ solutions in the field. This involves measuring water or air quality in remote areas. Still, more generally, we can think of these as standalone sensors that can be placed anywhere there is a need for measurements. The 6Ps govern the basics of environmental monitoring. These are the results of the standard 5W1H analysis for a scientific problem. These Ps consist of Premise, Purpose, Process, Place, Persona, and Period:

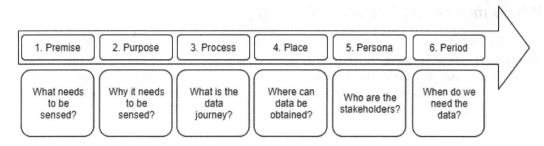

Figure 3.3 – The 6Ps to designing environmental monitoring systems

Using these items as guidelines can be a good exercise when you're looking to implement a standalone sensor strategy. *Figure 3.3* defines the following:

- **Premise**: The premise forms the first important question that needs to be answered during this exercise. Premise establishes the context of the functional requirement. It defines all the parameters that need to be identified and measured to calculate specific KPIs to solve them.

- **Purpose**: Sensors are the means to solving a problem. But they do not become the solution. Identifying the reason for sensing a particular parameter is vital to avoiding over-engineering and managing your budget.

- **Process**: Process consists of the choice of technology, the selection of protocols, the sensitivity, the precision, and so on that results from installing and commissioning the monitoring system.

- **Place**: There are usually multiple data sources and, more importantly, heterogeneous ones. The next step is identifying the optimal tap points for data availability or installing retrofit sensors.

- **Persona**: The persona or end user is also a critical design factor. Depending on the usability, importance, and applicability, the choice varies. As part of the shopfloor entry system, a biometric can be designed based on face or speech recognition. But if we consider speech-disabled users, the bias toward face recognition increases.

- **Period**: Data acquisition frequency also plays a critical role in designing a solution. A balance must be reached based on how often we need data to be acquired. This is usually arrived at with a combination of how drastic the values change and how massive the impact is on the system.

All this provides a way to put a better strategy around your choice and deployment of sensors. Understanding what, how, and why you are deploying a sensor strategy is essential to ensure it meets the needs of the business. It also helps provide a roadmap for which type of sensor you wish to use and the frequency and velocity of data from various systems.

There is a seventh P, which is usually **price** or cost. A critical factor is profit drives the business. However, this has not been included as part of the 6Ps to separate financial decisions from technical and functional requirements. There may be tradeoffs later in the type of sensors or networks you use, which can influence your cost/benefit analysis efforts when making decisions.

Why is monitoring important?

A famous quote states, *"you cannot manage what you cannot measure, let alone improve it."* So, measuring and monitoring are the foremost principles of any system. The industrial manufacturing ecosystem is no exception to this. The Six Sigma data-driven quality strategy for process improvement has measure/monitor as the core component, and after every stage, the direction and results are measured for decision-making:

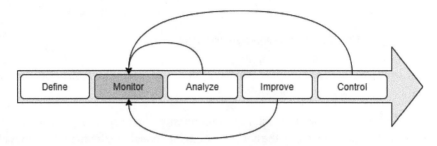

Figure 3.4 – The Six Sigma core model

We understand that monitoring at the site is critical. It is pivotal that we measure in real time and capture essential metrics and variations. We need to then analyze and transmit this data northbound as soon as possible and ensure we have as much security in place toward the decision support application layer. *Figure 3.4* provides a simple view of how sensor deployment, data analysis, and control systems can work together to provide substantial business value.

We have now looked at various sensor types that aid in real-time monitoring, and we have also looked at contextualizing various aspects of sensors through 6Ps. Now, let's briefly delve into integration patterns and popular protocols that help in data transmission within a heterogeneous environment. Patterns and protocols can be visualized as structured templates and digital language translators that help make communication easier and more standardized.

Integration patterns and protocols

Integration patterns and protocols are the segway for data transmission across the layers. As an industry, we have expanded the original OSI model to suit our evolving IoT needs. This model consists of four essential layers within which multiple sublayers are abstracted – the hardware layer, the infrastructure layer, the management layer, and the business layer. Roughly, the infrastructure and a part of the management layer are part of the on-premises setup or the edge setup. The importance of protocols across layers is visualized in *Figure 3.5*:

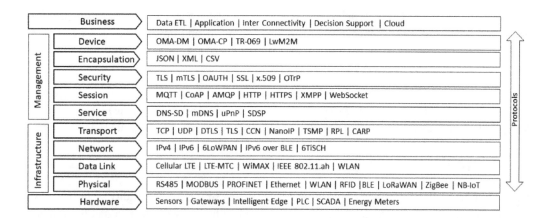

Figure 3.5 – Protocol stack across the four layers

The hardware layer consists of the OT technology components that interact with the production process. There are the sensors and PLCs that interface and run the machines, SCADA systems that perform data-based control, energy meters, connected gateways that talk multiple protocols, and the intelligent edge, which is capable of acquiring, storing, pre-processing, and transmitting data to the cloud. The hardware layer is enabled by many industrial automation protocols such as CANopen, NMT, PDO, SDO, and others, which we will discuss in future chapters in more detail.

The infrastructure layer provides a medium for transmitting the acquired data from physical entities to the cloud. There are wired and wireless mediums for data communication, and each evolves based on bandwidth, speed, reliability, cost of operation, battery life, range of distance, networking capability, and more. It encompasses the OSI data model's physical, data link, network, and transport layers. The transport layer still heavily uses TCP, UDP, or modified forms of these to send data to the others layers.

The management layer deals with the nitty-gritty of the governance of data transmission. Starting with device management, authentication, and security, it also handles session controls and data packet encapsulation and formatting. Multiple major and famous protocols are referenced in the diagram. This layer creates the rules for safe data transport to the business layer.

While the business layer can still exist on-premise at the edge (in an on-premise data server), within our model it is conveniently placed in the cloud for scalability, high availability, and expansibility across the enterprise. It's comprised of the data ETL, which is responsible for extracting, transforming, and loading data, which is the primordial step in transforming multiple data streams into useful information. It also contains the data lake we discussed in the previous chapter for storing data. The interconnecting layer also pulls data from other business and enterprise systems like ERP. The application layer, which is built on intelligent algorithms, creates cross-computation and correlates data. Along with providing historical inferences, this helps create a knowledge map. This super helpful knowledge graph is then presented to various stakeholders or personas through visualization applications. Then, decisions are recommended, or anomalies are identified, and actions are highlighted.

Thus, the decision support system helps fine-tune the physical systems and processes to aim for efficient operations and performance improvements, thus minimizing waste, losses, and downtime. While the various protocols are parallel in terms of functionality, their interoperability and aptness become a choice of design, existing ecosystem constraints, and hardware.

We now have a basic grip on the methods for obtaining data. The next step is to immerse ourselves in the means for getting that data. This process is not always straightforward, given the highly complex, heterogenous, multi-aged, and vendor-specific ecosystem we see in every brownfield implementation.

Getting data from the field

Data acquisition is one of the most challenging aspects of any solution. The validity and usefulness of all applications depend on the timeliness, accuracy, and authenticity of the data that fuels them. Given both brownfield and greenfield scenarios, lack of standardization, multiple system integrators, and the heavy usage of OEM-specific proprietary protocols, there have been many complex additions to the data acquisition problem.

OEMs have been reluctant to open data ports, beginning with the guarded ecosystems, sighting security, and warranty issues. This can prove to be a bottleneck for accessing critical operational data. Usually, production lines are revamped one at a time over the years, implying that there might be multiple generations of machines from one or multiple vendors working on evolved and legacy protocols. This results in a completely heterogeneous ecosystem. So, there cannot be a one-size-fits-all solution in many of these cases.

We also have systemic inhibition from the workforce in a top-down-driven system where misconceptions based on automation often result in redundant jobs. This is a significant change management topic reserved for a book in itself.

Since the idea here is to acquire data from multiple sources in a time-bound, error-free, secure, and accurate manner to create a unified data model, we work on the ideology of in-situ environment monitoring. This process broadly consists of defining the functional business problem of what needs to be measured and how measuring that would steer the system toward improved operational efficiency. After analyzing the possible open ports, APIs, system databases, and local automation systems, we also evaluate the possibility of retrofitting additional sensors to acquire more data. The integration patterns and protocols are thus dependent on existing systems or are unified based on architectural tenets. The result is to have a working pipeline streaming data from the production line to the cloud, where the decision support system takes over:

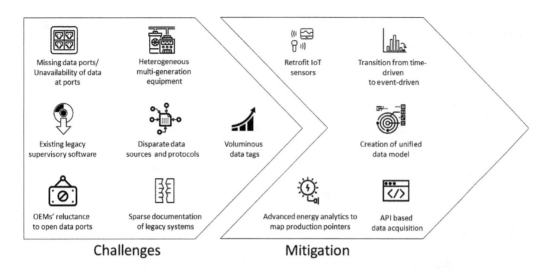

Figure 3.6 – Data acquisition challenges and mitigation approaches

Data drives everything – a common use case

This section will review two fascinating use cases of in-situ environment monitoring and their surprisingly complementary outcomes. While this section talks about the technical implementation of the solution, a real-world example using AWS will be presented more fully in the next chapter.

OEE for a legacy plastic manufacturer

The following use case belongs to an injection molding machine manufacturing line. This is a legacy machine that was commissioned decades ago. While it is still functioning, there are a lot of inefficiencies due to downtime, maintenance, and so on that are impacting the outcome of finished goods. The existing setup consisted of a machine that produced molded plastic parts continuously, with inputs being the shaping mold and raw plastic sheets. The produced goods are manually checked, and quality is passed to the packaging department:

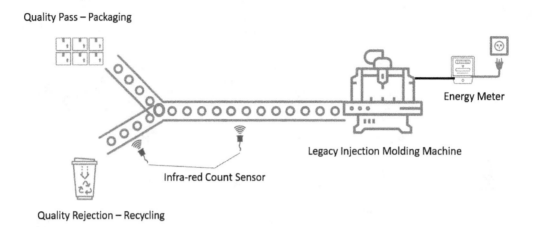

Figure 3.7 – Injection molding production line and in-situ monitoring

In the existing environment, with the introduction of three simple, low-cost sensors, the production line is now transformed into a measurable more intelligent system. Let's see how.

An energy meter interfaced to the local edge via Modbus RS-485 provides vital electrical statistics such as current drawn, voltage, power, energy consumed, power factor, and more. By adding infra-red count sensors on the main output line and the quality rejection line, we can provide accurate measurements of the total produced and total quality rejected parts.

Overall Equipment Effectiveness (OEE) is one of the defining parameters of working efficiency, be it at the machine level, line level, or plant level. OEE is the product of availability, performance, and quality:

Overall Equipment Effectiveness = Availability x Performance x Quality

Availability = Run Time / Planned Production Time

Performance = (Ideal Cycle Time x Total Count) / Run Time

Quality = Good Count / Total Count

Figure 3.8 – Overall equipment effectiveness

Availability is the *actual machine run time* ratio versus the *planned production time* for the day. The actual runtime of the machine is now directly derived from the energy meter, where a counter increments as and when the current drawn for operations is lower than the setpoint for normal operations. We also have the two infrared-based sensors' total count and good count values. With ideal cycle time already keyed in as part of standard averages, we can efficiently compute OEE and thus monitor the entire production line.

> **Understanding normal operating efficiencies**
>
> An interesting outcome of this simple example could be reflected in our employees' daily behavioral patterns. For example, an hour before lunch, there is a dip (hunger, maybe?) in production, resulting in more manual machine stoppages. Before the close of the shift, there is peak efficiency to meet the daily target. This essential in-situ monitoring exercise of making a legacy production line a little smarter helps us visualize critical operational metrics, assists supervisors in setting realistic production targets through historical averages, and perhaps gamification, the process of improved production efficiency and reducing unplanned downtime.

OK, we realize this is not the most realistic example, but it is simple enough to provide an idea of the approach. We are explicitly looking at productivity and downtime rates in this example. Still, this chapter and the next have a much broader purpose: providing you with some ideas of where monitoring makes sense and how easily it can be accomplished.

Adventures of a humble data packet

In most process-based industries dealing with chemicals, treating effluents within the factory premise is one of the critical steps toward following government laws and regulations. The effluent treatment removes contaminants from wastewater (by-products of manufacturing, cleaning, and so on), including physical, chemical, and biological processes producing environmentally safe treated wastewater (or treated effluent). Therefore, sewage, effluent, and wastewater treatment are critical for the compliance and business continuity of the entire plant. A failed **Effluent Treatment Plant (ETP)** can halt whole production lines. Thus, continued operations of the treatment plants are highly critical for business continuity

But the non-availability of a workforce to operate ETPs, expensive maintenance costs, super expensive failure/stoppage costs, and higher operating costs at lower efficiencies are some of the fundamental issues plaguing these systems worldwide. This calls for a classic in-situ environment (in a literal sense) monitoring system, which would detect operational anomalies in near real-time and help plan maintenance without disturbing actual production on the shop floor.

The monitoring system consists of multiple sensor units strategically positioned to measure specific parameters. Similar sensors are also distributed to create an average net sum of the measuring parameter. There are derived parameters such as **Biological Oxygen Demand (BOD)** and **Chemical Oxygen Demand (COD)**, computed from raw sensor values. While the variable frequency drives provide a snapshot of the motor performance, correlating with historical energy consumption for a given flow rate provides values for operational efficiencies. There is also a local on-premise control system for dousing chlorine as per the recorded real-time values.

A seemingly simple aerobic aeration by bacteria to break down biological and chemical waste has more than 40 sensors to determine and optimize the future state accurately. How does this type of monitoring map to our architecture?

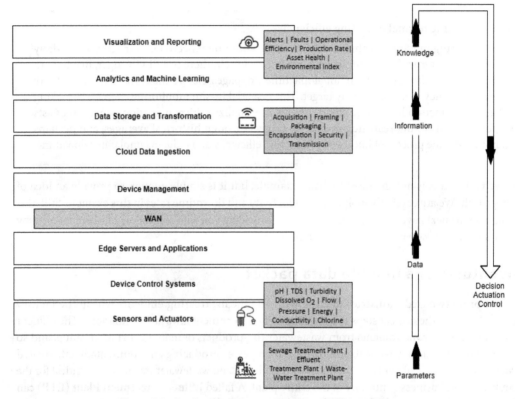

Figure 3.9 – In-situ condition monitoring of ETP in a factory

The full-time data-driven visibility ensures consistent operation with minimal energy consumption and provides a planning window to time maintenance changes with suppliers and in-house personnel.

Within the plant, processes are in place to help identify and fix these problems. But what about outside the plant? Runoff in fields and streams needs to be measured at the source and possibly kilometers downstream to determine the effect of plant behavior. Additionally, there may be other sources, such as agricultural runoff from spraying the fields or contamination from cattle and ranches. Environmental monitoring can play a much more significant role in these areas.

After this thrilling deep dive into the data packet that we have undertaken, it is time for the million-dollar (or many thousands) question of digitization: wired or wireless?

Wired versus wireless

Wired sensors have some obvious advantages. With a hardwired connection, data can be transmitted much faster and without delays over short distances. Routing power to the sensor is also not a problem; most industrial sensors that are being used to collect essential data from machine networks should probably be hardwired connections. This is mission-critical data and collecting and analyzing that information without interruption is also critical. However, wiring and connecting those components can come at a high cost. When the signal traverse length is extended due to distributed and decentralized peripheries, signal loss, and speed is possible. Here, wireless technology with radio transceivers gains speed range and bandwidth advantage.

In recent years, say the last 6 to 8 years, wireless sensors have become more prominent in manufacturing and environmental monitoring. Advances in technology and low-cost sensors have been a significant driver in this effort and the ability to place sensors pretty much anywhere. But it is not all sunshine and rainbows – in industry, there are usually very rigorous standards for what goes on on the factory floor. A two-dollar temperature sensor, driven by an off-the-shelf Raspberry PI, is not generally ready for the rigors of an industrial setting.

Wireless in-situ monitoring

The authors have completed dozens of IoT projects in different industries, such as agriculture, water quality, and energy sectors, such as oil and gas. One relevant early-on project was for a large food manufacturer in the midwestern region of the US. The manufacturer had been working with other vendors to try and build a sensor solution for their factories with limited success. The use cases were broad and varied and didn't necessarily apply directly to the manufacturing line. Still, they would be valuable to the other process areas outlined in *Figure 3.1*. The cost was a factor as running new cable for a wired solution would be tens of thousands of dollars per line, and for some of the use cases, this was not possible.

Our solution was to place three LoRaWan gateways in different plant areas to try and get the broadest coverage possible at a relatively low cost. The result was terrific. For example, we placed parking and door sensors at each bay door in the warehouses, which allowed us to see which bays were available and which bay doors were open (a potential safety hazard). During the signal analysis, we noted that the signal to the nearest gateway was powerful, even with the sensors placed under the asphalt. What was more surprising was that we also received a signal from the farthest gateway, which was well over a mile away, and traveling through or around tons of concrete and steel to deliver that message.

Granted, that message was a single value, a one, or a zero when something parked over that sensor. The benefit of a tiny message with a long-range wireless protocol allows that type of monitoring quite easily. Once the gateways were installed, it became easy to leverage the available network with additional sensors for monitoring temperature in network cabinets or outdoor cold storage trailers.

In addition, there are safety concerns in many industrial situations – for example, the use of explosion-proof cases so that the electrical components do not ignite free gases or combustible dust particles. The waterproof and dustproof casing is also required to withstand the rigors of dirty or exposed locations. **Electromagnetic interference (EMI)** is also crucial, especially in densely connected component layouts. Unregulated EMI designs without EMC can lead to signal losses at a minimum and can also lead to physical system damage. Finally, weather can be a factor, where sensors stop working in freezing temperatures as the silicone freezes and batteries stop working. With wired sensors, this can be less of a problem if power is applied from an external source, and internal heating can be used to maintain temperature requirements for the electronics.

Even with these constraints, things are catching up, and cheaper and better industrial-ready solutions are now more readily available, with new ones released all the time.

Wireless network evaluation

As with all things in the technology realm, there are choices, such as considering which vendor, protocol, or wireless network to use. In previous chapters, we discussed a little bit about some of these considerations and decision points. *Figure 3.10* provides a simple view of some of the most widely used wireless networks and differences in range and data rate. Power consumption is included since a higher data rate equates to higher power usage in most cases.

Standardization is important. We want to standardize our sensors, networks, and data systems to reduce or at least contain our total cost of ownership. Mixing and matching will cost more in implementation and maintenance if consistency is not maintained. Along with that, some considerations for the network (and all your edge devices and systems) include the following:

- **Power**: Power is sometimes an afterthought, but several factors should be considered. Range and bandwidth determine how much power you might need. The frequency of data being sent is also an essential factor. If you send data frequently, you need a power supply that can maintain constant usage. Along with use comes the size of your power source. A couple of C or D batteries will not last long under heavy usage and high bandwidth, even if they may work very well (several months or years) with lower frequencies and low bandwidth messages. Finally, consider rechargeable options, such as solar, to maintain your power cells at a premium level.

- **Range**: **Low Power Wide Area Network (LPWAN)** options such as LoRa are great options for long-range. With that range comes lower bandwidth restrictions. Since some protocols are unlicensed, there are restrictions on the number and size of messages that can be sent within a specific period. But for some sensors, which are only sending short messages at longer time intervals, LoRa can be a great choice.

- **Bandwidth**: Are your bandwidth needs high? Most in-situ bandwidth needs are low enough for this approach to be considered. This may include sending a single on/off type of value or an encoded string of water quality measurements. If you send images, for example, or very frequent data, wired solutions with consistent power and high bandwidth capability are possibly a better solution.

- **Availability**: Is there a network endpoint available within the range of your devices, or will you have to deploy endpoints as part of your engagement? Additionally, are sensors and sensor nodes available to take advantage of that specific network? We mentioned this potential drawback of the unavailability of matching sensors in *Chapter 1*.

- **Cost**: The network and data transfer prices are important to consider. Sometimes, it is hard to evaluate how much traffic you will send through the system; however, it is necessary to initially try and estimate network costs at scale. Consider that many (millions) small data packets can be expensive to transfer for most cloud providers.

Mixing and matching technologies

A couple of cases come to mind where combining wireless approaches has been necessary. Both happen in the oil and gas industry, where safety around hazardous materials is prioritized. I will outline one case that required the instrumentation of saltwater disposal wells used to dispose of water after a fracking operation. These sites usually consist of multiple tanks where water is trucked and deposited, and any leftover oil is separated and stored before the water is pumped down a well.

Our approach used an explosion-proof, industrial-depth sensor inserted into a hole (knockout) at the top of each tank. Floats riding on a cable would measure the depth of the liquid at any given time, and the sensor node (RTU) would transmit the level to a gateway for measurements. These floats were interesting as they had specific gravity densities for the type of liquid they were floating in. The wireless protocol, however, was proprietary. To accommodate this, we adapted a Modbus RTU adaptor with a 4G backhaul to read and send the data to our cloud-based system for monitoring.

It feels pretty simple when you understand how the solution was implemented; however, at the time, we were at the beginning of an IoT revolution, and putting the pieces together was quite innovative and challenging as the market was too slowly evolving to meet our needs.

Now that we have understood the nuances of evaluating a network based on the parameters discussed above, it is time to choose one based on the solution requirement. Please remember that there is no one-size-fits-all solution or a good or a wrong solution. Again, the answer combines conditions, choices, options, and context.

What network should you choose?

It takes years for a wireless network to become ubiquitous, and some never really see the light of day. We can name at least a couple that has gone the way of the dinosaur not that long ago. Others are very proprietary. Use our network, protocol, gateways, and sensors; everything will work together great! Sometimes, that can be the correct answer and reduce a project's complexity. However, it may only work in a small number of use cases. It is common to hear from these vendors, "we don't have a sensor for that."

Figure 3.10 outlines a simple comparison of some of the different prevalent networks available and where they fit in terms of range, bandwidth, and power needs for the devices sending data.

Over the last decade, we have used all these protocols and technologies in different situations. Each one is considered independently of other customers, location types, and the types of measurements being gathered.

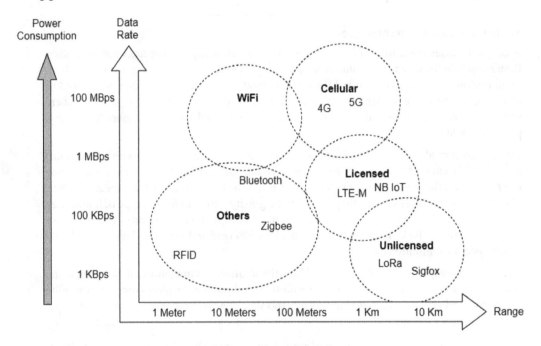

3.10 – Wireless network comparison

We can identify the significant differences here but be aware that this figure does not consider every available technology, nor does it place specific vendors within each technology or with their technology or protocol. Let's look at the major protocols and their areas of application:

- **Wi-Fi**: Wi-Fi has been around forever, it seems, and sometimes can be a good approach. It can be costly in terms of power consumption and bandwidth, but when sending large messages such as images it provides a cheap backhaul of those images to the cloud.

- **Cellular**: Cellular networks such as 4G and 5G can be a great fit if you are in an environment where other options do not exist or you cannot set up a wireless gateway or receiver. Urban environments make sense here, where the cost of a small number of sensors communicating directly with the cellular network outweighs the cost of setting up a gateway for a different protocol. So long as you have cellular service in the area, it can work quite well. However, be cautious of power consumption on your sensors, as connecting to a cell network can take a hit on your battery if you do this too frequently. Cellular communication is often used for backhauling your data to the cloud from the gateway, which then talks to the sensors using a different protocol, such as via an LPWAN approach.

- **Licensed LPWAN**: Licensed and unlicensed LPWAN is where our primary focus for this effort will be the focus of the next chapter. The most significant difference is the frequency with that these protocols use. Licensed LPWAN primarily consists of **Long-Term Evolution for Machines (LTE-M)** and **Narrowband Internet of Things (NB-IoT)**. Both protocols are becoming more widely available, but not consistently. Hence, it is important to determine which carrier is deploying the networks in areas where you require access. Cost and bandwidth are lower than cellular but probably a little higher than unlicensed LPWAN, depending on your options. This can be a good option if the network is available where you need it, especially in urban areas.

- **Unlicensed LPWAN**: The unlicensed spectrum has become a major player in the IoT industry, with LoRa and Sigfox driving most of the heavy lifting. Some proprietary networks and protocols also operate in this spectrum, many of which are solution-based and provide end-to-end hardware/software for monitoring a particular area. The former took slightly different approaches by allowing end users to leverage their protocol or network more openly. This allows for much more flexibility and availability of solutions for different monitoring situations. We will delve more into LPWAN as a whole in the next chapter.

- **Others**: We kind of lumped everything else into this group. This includes Bluetooth, Zigbee, and even RFID. All of these have their uses, but they will not be a strong focus of this book. Approaches such as RFID or Bluetooth can be used successfully in industries that need features such as track and trace. These low-cost and low-range devices allow you to track everything from widgets to pallets to entire trucks as they pass in and out of the warehouse or loading dock. Maybe that should be the focus of the next book.

People counting with Wi-Fi and Raspberry Pi

Wi-Fi, in combination with a Raspberry Pi, hardly seems like an industrial solution. We could find some use cases on the factory floor, but they can be useful in some situations. Early in the IoT revolution, we worked with a medium Texas restaurant chain. Our initial objective was freezer monitoring and alerting when the temperature was too high. It was a common practice for the teams to leave the freezer open for a few hours bringing food in and out of the freezer as workers prepped for lunch or dinner. While not unusual, there was a concern it could lead to spoilage if not monitored closely.

One of the restaurants was in a trendy area with a lot of foot traffic and many competing stores and restaurants. As we looked for more use cases to add value to the restaurant, the manager asked if there was a way we could determine traffic patterns and how the restaurant could attract more customers. The store could tell volume based on sales receipts, but could we track its face marketing and see how it might affect customer volume? Our approach was twofold.

First, we determined traffic and flow outside the restaurant. We accomplished this with cell phone monitors placed at either end of the restaurant to capture the number of cell phones in the area and the direction of travel as they moved. Additionally, we could see how long these phones were in the area, presumably in or around the restaurant. This gave us an idea of foot traffic volume and flow at different times of the day or when there were significant events in the area, which was often the case.

Second, we wanted to count how many people came into the restaurant. We developed a camera solution on a Raspberry Pi that overlooked the restaurant's foyer and counted heads as they entered or exited the front door. Really, with current technology, this is nothing more than a high school science project, but at the time, it was an innovative and cost-effective solution to a tricky problem.

Later, we used some of these same cell-phone- and people-counting solutions in other high-profile areas, such as with the mobile signs on the Las Vegas strip, to determine how many impressions may occur as the signs passed through key areas.

We have provided this gentle introduction to wireless networks and protocols to help introduce you to the concepts and options available. If you are already familiar with some of this, then you know how things work. We purposely avoided details about frequencies and durations and technical information about how these protocols work. It seems unnecessary to bog you down with too much information that can be looked up online.

The following section deals with one of the most sought-after questions in the solution stage of an Industry 4.0 digital transformation project. We will discuss wireless gateways and intelligent edge devices, what they are, and which ones to choose for a given use case.

Wireless gateways versus Intelligent edge devices

The gateways we will deploy in the next chapter are fundamental because they are set up primarily for message forwarding. They do not have any real intelligence other than to forward a message to the cloud and possibly wait for a response. The response is then sent back to the device as an acknowledgment or may be sent as a command.

This is fine to start. We want to take this in baby steps. However, as data volume grows and becomes more critical, we will look for more ways to process, pre-process, or even act on data as close to the device or location. We call this the edge –the place in the factory between the machines before data gets sent to the cloud.

This may be one drawback of commercial or third-party wireless networks. The data often goes directly to the cloud for processing, missing the opportunity to perform work closer to the system (the edge) and react. In most cases, this is not a problem. The monitoring that we are discussing is a relatively small volume and small packet size. But this should be something to consider if you think your need for processing at the edge is more critical.

In this section, we will discuss the edge's importance and evolution. The edge has been theoretically defined as the farthest compute unit away from the cloud. It was usually seen as the connecting piece from the shop floor to the top floor (from production machines to the cloud). It is quintessentially responsible for acquiring data from many devices, speaking multiple protocols, packaging them, and then transmitting them to the cloud. This basic acquisition and transmission functionality device is called an IoT gateway.

An IoT gateway is a communication-capable device responsible for receiving data transmitted by IoT devices from various locations and feeding it to the network server for further processing. IoT gateways are powered by microprocessors running firmware and real-time operating systems. They also have hardware integration points for data acquisition. Usually, gateways are not designed to process, filter, or store data. Being smaller footprints in terms of memory and computing, they forward data to an edge or the cloud directly.

IoT gateways have evolved from mere data aggregators and transmitters to sometimes powerful computing devices. These are now known as intelligent edge machines. While acquiring and communicating data, edge machines can store, compute, and actuate locally. They can deliver commands from the cloud application to the end devices. *Figure 3.11* compares the IoT gateway and the intelligent edge machine. The term intelligent edge machine is not to be confused with production machines (robots, CNC, etc.) with built-in intelligence. We will discuss devices with intelligence and M2M communication later in this book:

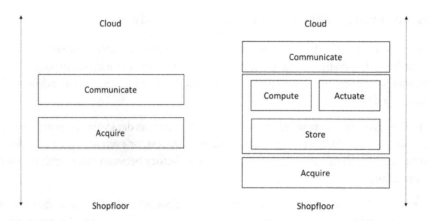

Figure 3.11 – Evolution from IoT gateways to intelligent edge machines

The anatomy and the various layers of an intelligent edge can be seen in *Figure 3.11*. The layers between the hardware sensor layer and the cloud layer are collectively considered the edge and will be described in the following section in detail. In the upcoming chapters, we look at the exact data flow to and from each layer within the context of real examples.

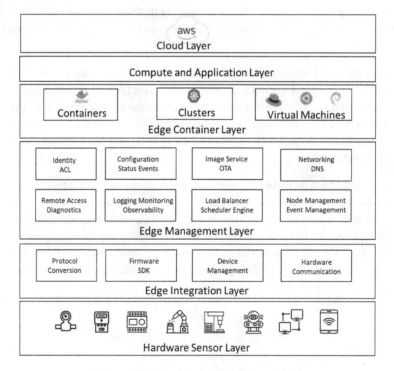

Figure 3.12 – Layers of an intelligent edge

Various layers potentially form the edge environment. These layers potentially take care of data acquisition through multi-protocol conversion, local compute, storage, and secure transport to the cloud:

- **Edge integration layer**: The edge integration layer handles various protocols via firmware and SDKs. It also has different hardware ports for data acquisition and transfer. Device management in storage and special services and drivers also belong here. For example, it houses drivers for the BACnet, MODBUS, and RS485 protocols and has a stack for WLAN radio, Zigbee, LoRaWAN, and others.

- **Edge management layer**: This layer has evolved for autonomous and scaled handling of multiple networked edges. Edge management deals with the efficient functioning of the device(s), focusing on collecting operational data. These can be the observability layer's metrics, events, logs and traces, presented to the user as alarms and notifications. Self-run diagnostics also trigger through algorithms that feed off this data. Security is managed through identity and access, which validate and secures the device within the network. We have the networking, load balancing, and traffic routing services built as part of this layer. Configuration management is also an important service that accounts for running the right and latest versions of the applications. The OTA and image service downloads and pushes firmware updates from the cloud. These updates could be for applications running on the edge, drivers responsible for data acquisition, or the firmware of the end devices that are interfaced. The scheduler engine can run pre-determined cycles of commands on the edge or pass them to the end devices based on the outcome of algorithms at the edge or directional input from the application in the cloud with human intervention. But considering security issues, automatic remote command execution for critical machine operations is usually not recommended.

- **Edge container layer**: The edge container layer manages various workloads that run as docker images, Kubernetes, or virtual machines. Each runs specific applications that consume the data obtained from the lower layers and affect intelligent decisions. Managing system resources such as storage and compute and helping in inter-pod communication are major responsibilities of this layer. The compute and application functions are usually part of this layer. It has algorithms that help in localized decision-making in terms of network unavailability or by design. The applications also have the medium to communicate or send data or processed information to the cloud counterpart, where detailed analytics and machine learning models are run. Machine learning models and inference engines can also be run locally based on the hardware capability chosen.

There is potentially a lot of complexity for computing at the edge. This is just an overview of what could be in store. The answer is always to take it slow or use systems with built-in capabilities to manage this complexity.

Summary

In this chapter, we discussed in-situ or onsite environment monitoring and its purpose and importance in the bigger picture of an intelligent enterprise. We then undertook a journey into the means of their implementation through sensors, types, and selection criteria. The current integration ecosystem and protocols were also discussed while foraying into two interesting industrial use cases that presented a solid case for monitoring.

In summary, we discussed the fundamentals of the DIKW model, from which humble data transforms itself into intelligence, enabling stakeholders to optimize their systems efficiently. Data is a set of random, seemingly non-correlating signals from physical systems. When given context, data becomes structured, organized, and valuable information. When given meaning or context, information becomes knowledge that reveals patterns and synthesized behaviors from history. When armed with more insight, knowledge becomes intelligence or wisdom, which is actionable and integrated. This can alter the future for the better. Finally, intelligence becomes decisions when given purpose and the change that needs to happen as systems evolve constantly.

We have been through a lot of theories and ideas so far. The next chapter will focus on simple real-world monitoring examples using industrial hardware and getting data to the cloud.

4

Real-World Environmental Monitoring

We are excited about this chapter because we will finally put words into action. The previous three chapters included much discussion and theory about how to approach things when designing your architecture and some of the considerations in getting your initiative started. In this chapter, we will focus on hands-on environmental monitoring, which has slightly different goals than what we might initially consider for industrial monitoring in a manufacturing environment.

For some industries, environmental monitoring may be your primary goal. Agriculture and ranching, for example, are widely diverse industries that depend significantly on environmental factors. Oil and gas, mining, and maritime sectors also have use cases in this area. Even traditional manufacturing needs environmental monitoring that can support the manufacturing effort. This could include cold chain tracking or monitoring areas that can get overheated. Water and air quality factors can also benefit from this type of approach.

In this chapter, we are going to cover the following topics:

- Wireless networks and protocols
- Gateway and network setup
- Sensor configuration and deployment
- Collecting data in the cloud
- Setting up a basic dashboard
- Putting the layers together

We will walk through configuring and deploying sensors in several areas using industrial-grade sensors and gateways. Using purpose-built sensors, you will learn how to configure, deploy and monitor sensor data. You will also understand how various sensors and sensor nodes can be determined based on their usage, location, and frequency of data collection.

Technical requirements

In this chapter, we will gather data, configure sample sensors, and build an initial architecture to process the data flow and present it to the end user. Our goal is to get data flowing cleanly and set up parts of the architecture to build on for later chapters. You will fall into one of two camps:

1. A technical and hands-on developer or architect will dig into the technology and can follow this chapter to help you get going. This chapter, or this book, is not a basic step-by-step guide. It provides architectural guidance and high-level flow with links, code samples, and many details when necessary. However, if you fit into this camp, you will likely want to adjust things to your liking as you move forward.

2. Someone who is less technical will appreciate the lack of step-by-step detail as you can follow the approach and flow without getting bogged down (or maybe bored) in the details.

I think this approach provides a good mix of information. Honestly, providing too much detail would need several chapters or even books, and we are in a hurry to get to the next section and cover industrial data integration. You can follow the process and recreate the environment with some research and minimal background. You will leave this chapter with a good understanding of the basics of environmental monitoring and where to look for more information.

This chapter does require some additional hardware in the form of LoRaWAN sensors and gateways. You can recreate these examples and maximize your learning at the cost of a few hundred dollars.

You can find the code samples mentioned in this chapter at `https://github.com/ PacktPublishing/Industrial-IoT-for-Architects-and-Engineers/tree/ main/chapter04`.

Exploring wireless networks and protocols

Much of this book will focus on connecting to machine and industrial processes, but as we have discussed, ad hoc approaches for monitoring different areas of a plant are sometimes necessary. This chapter will walk through how to approach that goal. It will also allow us to define our architecture and understand how data is ingested, processed, and visualized. We will start by explaining the technology approach and then build some network components as examples.

In the previous chapter, we discussed wireless protocols and some of the pros and cons of various approaches. We would love to get more into the details, but we can't afford the page space. Most nitty-gritty details are widely available on the internet, such as the frequencies, range, and bandwidth of all the options discussed.

Standardizing your approach

In this chapter, we have chosen LoRaWAN as our wireless protocol example. Using LoRaWAN provides us with some benefits as our initial choice, and there are some points we consider when making our decision. Here are the challenges that network providers in the IoT industry face:

- If you offer a proprietary network, then it needs to be ubiquitous.

  The network has to be everywhere or at least very broad. Not just urban areas – we need industrial area coverage or the ability to expand the network relatively quickly and cost-effectively—at least one or the other, but hopefully both. You can tolerate a little more effort in getting things to work if it is inexpensive. You may expect to scale over time, and if scaling is going to blow the budget, it is a losing proposition from the start. The spectrum is the second theme, especially for the wireless, licensed and unlicensed.

- Compatible sensors should be widely available and varied if you offer a proprietary network.

  We have mentioned this a few times already. Proprietary network providers must invest in the availability of sensors and client devices. If you build it, people will not always come. People will come if there is a way to access the network quickly and as inexpensively as possible and access the data needed for their business case.

We also want to consider cost in this equation. As the trade-off protocols, such as 5G, become more of a factor and have more widely available sensors, making these decisions can be more challenging.

Architectural prototypes

At some point, you have to make some decisions and move forward. The analysis is good, but paralysis is not. We have all heard that over and over. In technology, the best approach is often to try something, especially when there are multiple reasonable options. When you hit a roadblock, it is not uncommon to go back and try something else. As a developer, this can be a very common refrain. Sometimes, after several different options, you come all the way around to your first idea again – developers know this process intimately.

In agile methodologies, we call this a spike, or an effort that lets you go deeper into prototyping a problem or solution without total commitment. It is a way to validate an approach. In any case, this could be an initial approach strategy when starting. Just set something up and test it in your environment. Remember that testing software is often cheaper than testing hardware. When getting started, we want to use a minimal budget for our hardware setup and minimize the investment without too many variations.

As mentioned previously, for our effort in this chapter, we will focus on the unlicensed spectrum, specifically LoRaWAN as the provider. There are several reasons why this is a good choice for our first attempt at using wireless IoT. A couple of them are as follows:

- LoRaWAN is an open protocol. There are many options for sensors, gateways, and network providers. This means you can purchase a sensor for any use case and have it work with this protocol.

- It provides a long-range, sometimes kilometers long, depending on the terrain. This means we can test many use cases with only one or two gateways.

There are many LoRaWAN network providers, so in this chapter, we will look at two different providers so that you can quickly see how things flow over various networks. The protocol is a little more transparent and easier to learn. There are also some negative issues using LoRaWAN, depending on your use cases:

- Bandwidth and use are theoretically capped for fair use. Since it uses the unlicensed spectrum, there are regulations for how much data you can send during a specific period. This is true for any protocol using an unlicensed spectrum.

- There are power transmission limits for both the uplink and downlink messages. You cannot send continuous data with the protocol, but you have limits on your transmission capability. It's powerful enough to be very useful and potentially has quite a long range, but it's not so powerful that you step on everything within your area of operations.

After some strong growth over the last few years, LoRaWAN is now widely used and one of the leaders in the unlicensed space, hence our decision to showcase it here. The requirements are pretty simple in concept. Of course, the devil is in the details, and all of those details would require a whole book and not a single chapter so we will leave them for another day. Generally, there is a specific flow of data in a system, which we have outlined in the following figure:

Sensor (environment) ➡ **Gateway** (strategically placed) ➡ **Network Server** (cloud based)

Figure 4.1 – High-level LoRaWAN message path

Sensors or end nodes capture and transmit the measurement values to locally available gateways. This sensor should be split into two main components, as we have mentioned previously: the sensor itself, which measures the actual value, and the node or transmitting unit, which communicates data between the sensor and the gateway. Often, these are one unit, but there are cases where they need to be separated within the environment. In the case of LoRaWAN specifically, these are small messages, usually encoded, are low power and long range, such as several kilometers or more in good conditions. Also, it should be noted that this data is infrequent in most cases. Messages can be sent every few minutes or hours, depending on the message size. This is generally the rule to ensure that the airways stay clear. LoRaWAN uses the unlicensed spectrum of 915 or 868 kHz, depending on your country or region.

Gateways receive messages or data from the sensors and then forward them to the network or cloud. Gateways are generally placed in locations with consistent power and a good connection to the cloud through Ethernet, Wi-Fi, or 4G/5G accessibility. This internet connection is called a **backhaul** and is used to deliver messages from the sensor, through the gateway, to the cloud.

Message forwarding protocol is the most common setup for a LoRaWAN gateway; however, there are cases where the network server resides on the gateway itself, and processing can be done at the gateway. However, this is not common, and if edge processing is critical, then perhaps LoRaWAN is not the correct protocol for your use.

Often, a sensor is in the range of several gateways and communicates the message all at once. In this case, the network server will receive the message multiple times, and depending on which gateway is closer or faster; it will decide which gateway should handle follow-up communication with the sensor.

The **network server** is the next step in the process and is usually a cloud-based system such as **AWS IoT Core** for LoRaWAN or **The Things Network (TTN)**. We will look at both of these systems, although dozens of options are available. This is where the real magic happens since the network server ensures that the device and message are valid and secure based on the device IDs and keys shared with the sensor node.

There are several types of message communications between the network server and the sensor node. Message types include join/accept messages, uplink and downlink messages to send data from the sensor to the cloud, and send a response back to the sensor if necessary. Downlink messages are generally configuration data specified by the vendor to change a setting in the node using an over-the-air configuration approach. This can be especially useful when sensors are out of reach, too numerous to update manually, or if the settings need to change frequently. An excellent example of this is when you need to change the frequency of sending messages from the sensor. As the measured values change, you may want to increase or decrease the interval of taking a measurement – for example, when irrigating crops. Irrigation can run for hours or days at a time, so long intervals are acceptable. However, when soil saturation starts to occur at the crop's root zone, requesting quicker data intervals and adjusting your irrigation effort could be essential.

Now, let's look at how we can set up our hardware and capture data from the environment.

Data capture architecture

We will take two different routes to get our data to the cloud, leveraging two various network providers along the way:

- First, we are going to look at **The Things Network** (TTN). This is a free and very popular option for our data. Once we have data flowing from our sensors to TTN, we will look at alternatives to transferring or forwarding that data to AWS for processing.

- Alternatively, we will send our data directly to AWS via the *IoT Core for LoRaWAN* service, a relatively new service in the AWS IoT family.

Why two? Primarily to illustrate some differences and similarities in how LoRaWAN hardware can be set up and transmit data. In both cases, our data will eventually end up in AWS, where we can examine, decode, transform and store it in our data lake. *Figure 4.2* outlines the two routes our data will take:

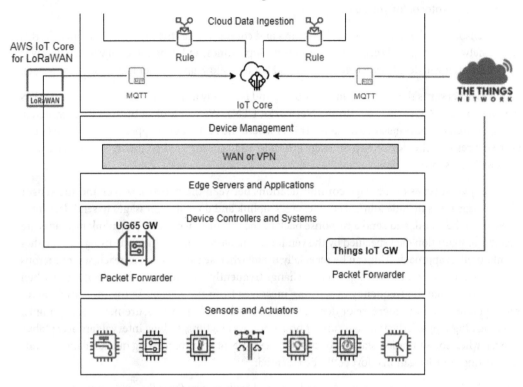

Figure 4.2 – Data capture and ingestion

Note that we bypass any concept of edge processing with this architecture. This is a critical point consistent with this protocol's low data volume approach. We could incorporate the edge by leveraging our network servers or placing the network server directly within the gateway, but this is a pretty advanced LoRaWAN setup that we will leave outside the scope of our example.

We will start by setting up the gateways and then look at some commercial sensors that can be used in a commercial or an industrial setting. The data from these sensors will then be processed and used later for basic monitoring and potential alerts.

Understanding gateway and network setup

Setting up the initial gateways is not difficult, and it seems to have become easier over the years. We have probably set up gateways with most of the network providers in the 900 MHz US space, primarily focused on the LoRaWAN protocol. Data from the sensor to the gateway and then to the network server is just one or two hops in our architecture. This is the act of getting your sensor data to the IT network or the cloud directly. From there, it can be identified and acted upon, along with other data streams.

We will try and stay focused on the overall setup and limit step-by-step details. Setting up a gateway will be well documented by your chosen vendor, and the details will change somewhat even before this book goes to print. The overall process has not changed much over the last few years, so this is where we will focus. We have chosen to highlight two providers in this section: TTN and AWS IoT Core for LoRaWAN. Specifically, we think these are good choices for learning and testing; however, we also believe they have long-term viability as providers in this space. So, let's get started.

Setting up a TTN IoT gateway

TTN has been around since 2015 and continues to advance in the LoRaWAN space. It was started and still runs as a community initiative; however, an enterprise component is available for business with the security, scalability, and support that most enterprises require. Our first gateway of choice is a straightforward option. We picked the TTN Indoor Gateway. It is easy to find on the web. If you google *TTN Indoor Gateway*, you will easily find a small white box with a built-in AC plug. It costs about 100 US dollars and has an internal antenna. With the built-in AC plug and internal web server for setup, it is probably one of the simplest ways to get started with LoRaWAN to set up an initial network that can grow with you if you decide to stick with TTN as a provider.

The process is very straightforward if you have acquired this gateway and are ready to get everything set up. It takes about three or four major steps to accomplish this goal, and nothing very technical. As a starting point for hands-on environmental or industrial monitoring, this is about as simple as it gets:

1. Connect to `https://www.thethingsnetwork.org/` to create an account with TTN.

   This first step is obvious. Create an account so that you can set up and manage your devices. You much also choose your region, either the US, EU, or Australia, before you can add gateways or sensors to the system.

2. Once you have your account with TTN set up, go to the console and choose **Gateways,** as shown in *Figure 4.3*. We have chosen the EU region in our case, so our gateway URL will be `https://eu1.cloud.thethings.network/console/gateways`. You will find that the TTN setup with this particular gateway is a little different. This specific gateway is built and sold to be used with the TTN network, so much of the configuration is already defined within the system. With third-party commercial gateways, you would normally add a new gateway to the system, but since this gateway is already defined by TTN, you need to claim it, instead of adding it as a new one:

Figure 4.3 – Claiming a gateway on TTN

This can be a great advantage when choosing a provider or service. If the gateways are preconfigured, then it can save considerable effort in the process. You are up and running by connecting and deploying the gateways in the field. However, this simplicity may come at the cost of some flexibility, so consider that when making your long-term decision.

3. Configure the gateway. The goal of configuring the device is to allow it to connect to your local Wi-Fi network to backhaul the messages that come from the sensor nodes.

    Configuring the gateway consists of resetting the devices to provide you with a local Wi-Fi network to which you can connect your computer. Once you have connected to the device's Wi-Fi and loaded the configuration page, you can choose or update your local Wi-Fi SSD and password to allow the gateway to communicate with the TTN network.

4. Verify that the gateway is connected using the TTN console.

    Once your gateway has restarted, you can go back to TTN and verify that it has been connected. *Figure 4.4* shows our device connected:

Figure 4.4 – TTN gateway connected

That is all for now. The gateway is connected and ready to receive data from a sensor device.

> **Making the network ubiquitous**
>
> At a TTN conference in Amsterdam several years ago, they initially released this simple hotspot type of gateway we are using here. TTN gave one to all conference attendees, which was some powerful swag for this particular group. It brought things closer to building a more global network since the attendees were worldwide.

But, in reality, a few thousand gateways (if they made their way to be configured) were only a drop in the bucket. After I got returned home, my gateway sat in the drawer for at least a few months, if not a year or two, until I got around to messing with it. By definition, no matter how much vendors try and sell you on it, LoRaWAN, or any unlicensed spectrum, is not as ubiquitous as they like to imagine, not like cell coverage and licensed networks are becoming today. You might get lucky, but more than likely, you will have to set up your gateways and define your network coverage, which, as you can see from this section, is relatively easy to do but could be costly for large areas.

We see this a lot. There is a lot of hype from network providers, claiming thousands or hundreds of thousands of gateways in the wild. I'm a bit skeptical myself and consider this useless information if they are not where I need them to be consistently.

With your gateway connected, there isn't much more to do now. Looking deeper into the TTN console, you can uncover connection messages sent from the gateway. This may provide a little more insight into your learning. We will look at the TTN console in more detail when we set up a sensor to communicate through the network and view the actual data messages being received.

There are dozens of LoRaWAN network providers across the globe, which is a good set of choices of which one you want to work with, different cost structures, and different benefits (such as mining crypto with your gateway). AWS has also stepped into the LoRaWAN ring with its network service. We think it is essential to introduce this latest AWS offering and review this as a capability and how it can be integrated into our overall architecture. For the AWS network setup, we use a Milesight IoT gateway, which provides several LoRaWAN gateways and sensors for commercial use.

Setting up Milesight UG65

Setting up the Milesight gateway on AWS is also reasonably straightforward. There are a lot of commercial LoRaWAN gateways on the market, and we have tried many of them for various projects. Milesight is a good fit for our examples since it is a proven commercial product with good flexibility and support. The version I chose was relatively inexpensive compared to others, and since we have a good relationship with the company, it was easy to send a few emails and purchase what I needed.

Previously, I mentioned that using LoRaWAN has become easier over the years. Most configuration setups for any gateway to a network provider are well documented. For our arrangement, which involves connecting the UG65 to AWS, a quick search has provided the following documentation: `https://resource.milesight-iot.com/milesight/document/ug65-quick-guide-en.pdf`.

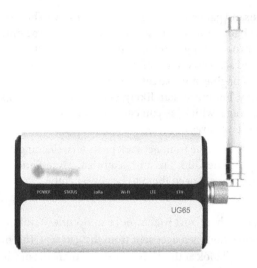

Figure 4.5 – Milesight UG65 gateway

The online guide walks you through the entire process of setting up AWS to connect and receive the data and setting up the gateway hardware and configuration. The UG65 can also connect to TTN as a packet forwarder and most other LoRaWAN provider networks. The steps are generally easy to follow, but some alignment may need to be performed to connect the gateway and AWS IoT. Here is a high-level summary of the steps:

1. Set up access to the gateway. This means applying power and connectivity, getting the IP address set up, or Wi-Fi attached. It is best to set up the gateway on the same network as your computer, where you also can access your AWS account. This will allow you to enter all correlating information into both systems to ensure a smooth connection.

    The UG65 provides several options for connectivity and power. It gives Wi-Fi with both client and server options and Ethernet and power or **Power over Ethernet (PoE)** options for flexibility. For me, choices often come with indecision and internal discussion about the best approach for the long term.

    I thought about options for a bit and finally decided to turn off the Wi-Fi on the device and use plain Ethernet and a power connector. Since this was initially sitting in my office, it was easy to connect and configure as I went through some examples. In the long term, I will switch it over to a PoE connection or possibly a power supply plus Wi-Fi connection, depending on where I permanently install the gateway.

2. Retrieve the gateway EUI. You will need the gateway EUI of the device. You can get this by logging into the gateway after you have set up the network connection. This value is the key to recognizing this gateway and is what you will need to enter into AWS to add a new device. In this particular model, I found it in the main menu under **Packet Forwarder | General**.

3. Add the gateway to AWS. Setting up an initial gateway in AWS is relatively easy. Many options exist, but we want to get data flowing since this is your first. Pick the basic options and leave that defaults as-is as much as possible:

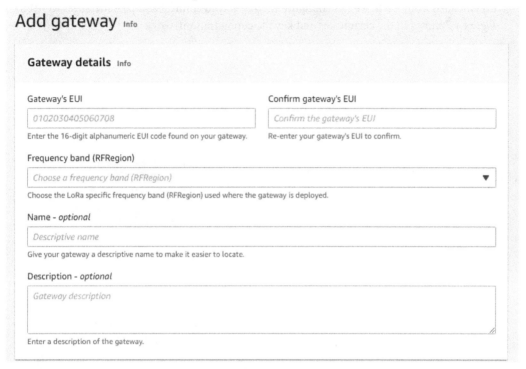

Figure 4.6 – Adding a new gateway to AWS

There is a little back and forth between the gateway and AWS. It would help if you had the gateway EUI from the device to enter into the AWS console.

In the AWS configuration, it is easy to get bogged down in the details of setting up a LoRaWAN device at this stage and try and understand all the nuances of setting up your network. My advice is to relax and have fun for a bit before we get too deep into the details, such as OTA versus ABP or determining which subchannels are the best fit for our use.

> **Important note**
>
> After you complete the add gateway step, *Step 1* in the onboarding process, you will be shown the next page, *Step 2*, to configure your gateway. On this page, you will need to generate a certificate for your gateway and copy all the provisioning credentials to your desktop. In addition, you need to download the server trust certificates and set up a role for the gateway within AWS. AWS can do this for you directly in this step. You need these files for the gateway setup so the AWS network can trust it.

4. Now, configure the UG65 to forward data to AWS. This final step is relatively simple. The UG65 has several operating modes and parameters based on your preferences. In our case, we are setting the gateway to simple forwarding mode. Using all the certificates and keys we downloaded from AWS, we can configure the gateway to connect and pass messages to AWS. *Figure 4.7* shows all the certificates and key file configurations:

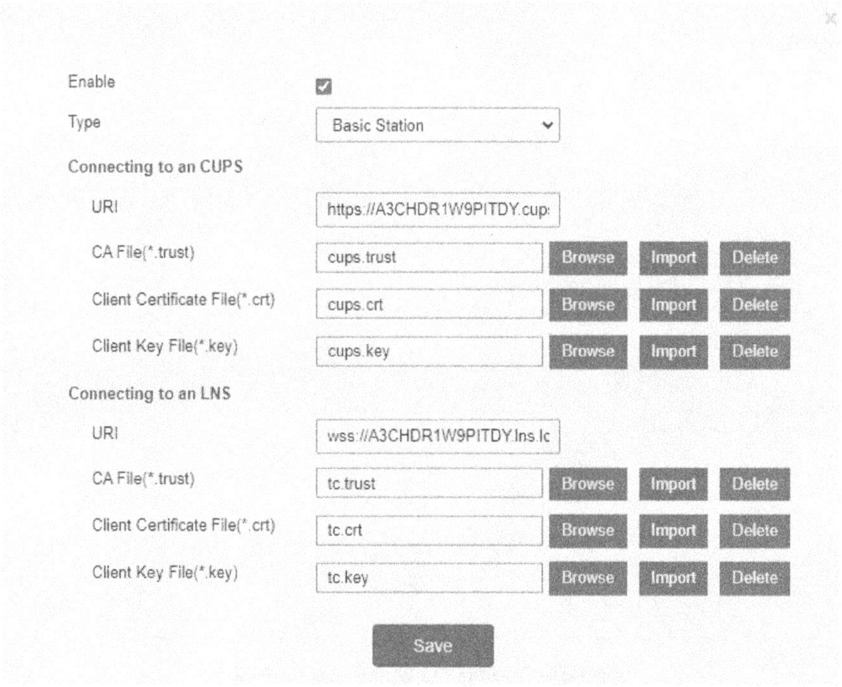

Figure 4.7 – UG65 packet forwarder

Basic Station or **Forwarding** mode is the simplest and most common way for a gateway to interact on a LoRaWAN network. The gateway forwards any messages (usually a join message) from the sensor and waits for a response. Multiple gateways will often receive and forward the same message since the sensor throws the message into the wild. In this case, the network server will determine which gateway should respond and receive the actual message traffic. The following JSON message from the device interaction shows the gateway that received the message from the sensor and how strong the connection is for each gateway in the list. In many cases, there will be multiple gateways in the area with different signal strengths, and surprisingly, some messages will go much further than you thought possible:

```
"Gateways": [
        {
            "GatewayEui": "24e124fffef239b8",
```

```
        "Rssi": -62,
        "Snr": 13.25
      }
    ],
```

Once the gateway has been configured and connected, it should start communicating with AWS and listening for messages from sensors within range. *Figure 4.8* shows the gateway configured and the last uplink received from the gateway:

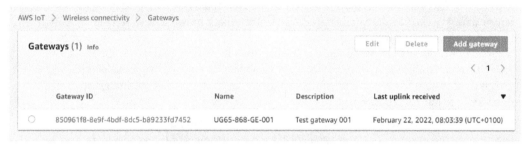

Figure 4.8 – Gateway connect validation

I wish we had more real estate pages to detail the different gateways and networks available today. There is a lot of research to be done if you are new to these technologies, but the study will help you make good decisions on your approach and configuration as you build out the network and sensors. We have just scratched the surface, but you can see that getting started and attaching some real-world equipment is simple enough. Both of these gateway setups are pretty cheap. A few hundred dollars for a few gateways and a handful of sensors can help you understand and get started regarding where this approach might bring value to your mission.

Now that we understand how to set up the gateways and network, it is time to review the setup of the sensors. Moving forward, we will configure a sensor to send data to the web to provide information about our environment.

Configuring and deploying sensors

In this section, we want data flowing from our sensors. Once we have a few sensors in place sending data, we can begin to use that to design and set up the cloud environment and think about how we want to process, store, and display the data. At this point, it is just raw data that we want to collect and visualize. Transforming that data into usable information comes in the later chapters as we think more about what the data means and how to combine it in valuable ways.

Cost and scale considerations

Several years ago, while working on a solution for the agriculture industry, our team was evaluating different wireless vendors and protocols for our solution. Sensors in the field mostly require long-range and low battery power to transmit ongoing crop, weather, and irrigation data for analysis and decision-making. We had about 60 locations to consider across hundreds of miles, which quickly grew to about 75, and we needed coverage for our sensors across many different farms, fields, and irrigation zones.

As you can imagine, cell coverage was spotty over acres of open field; however, our strategy was to use LoRaWAN gateway locations where 4G or Ethernet could be used as the backhaul for the sensor data. We evaluated several approaches for the sensor network, and several competing technologies were similar in the system's range, power, and scalability—one significant factor is cost. More propriety vendors offered to lease us the gateways at a price – of about ~50 USD each per month for one vendor. And that didn't include the network cost of accessing our data within their system. As you can imagine, this was a significant part of our budget to provide this service to our rural customers.

In reflection, this was yet another very prominent provider that has gone under over the last few years. In this example, there was no benefit for our customers or us to consider this option, especially when more cost-effective competing solutions were available. It is understandable that IoT network providers look to price their services at a premium but be sure you are getting value. It may require a considerable evaluation effort for large organizations that need to scale across a region or globally.

Now that our gateways are running, we can start sending data. The sensors we will configure are also from Milesight systems and represent almost any LoRaWAN-compatible sensor you can purchase from other vendors.

Connecting the AM319 multi-sensor

The Milesight AM319 is an ambiance monitoring device that provides nine measurement values. Many sensors in an industrial setting focus on one or maybe two measurements, such as temperature, pressure, or particulate levels. We have worked with these types of environmental monitors before, but what makes this one unique is the built-in display, which allows you to view the device directly as well as send the data for analysis:

Figure 4.9 – AM319 sensor

Setting up the AM319 is a two-step process similar to setting up the gateways: adding the sensor to AWS so that the network knows the sensor exists and configuring the sensor itself to enable a network join. It is easiest to do both at the same time, to configure the sensor, and at the same time, to add it to AWS IoT Core as a device. Before adding the sensor to AWS, a wireless device specification must be created. In our case, we used the **Over-the-Air Activation** (**OTAA**) 1.0 protocol for simplicity with all our sensors:

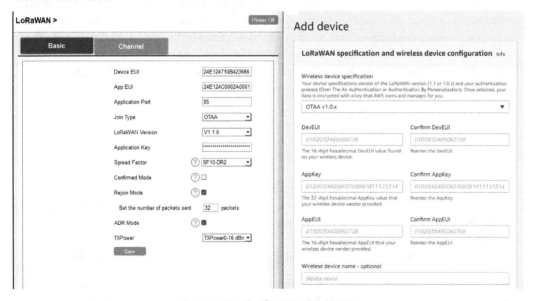

Figure 4.10 – Configuring the AM319

As shown side-by-side in *Figure 4.10*, common values must be configured in both devices. The DevEUI is key to ensuring a proper join can occur. But also, the **Application Key (AppKey)** and AppEUI are necessary to configure. Note that for AppKey, the default value was marked as * for security. You can also set up a new key for this integration.

On the AWS side, you need to create and provide a destination for your sensor. Generally, we would recommend starting with one destination for each type of sensor. *Destinations* tell IoT Core where to send the data from the device. This can be either an IoT rule or published directly to the IoT Core message broker to be picked up and processed by another service. If you decide to use an IoT rule, no topic name is required, but viewing your data on the message broker may be helpful as traffic flows in from your sensors.

Once these settings are configured and saved on the device and in the cloud, you should have data streaming to IoT Core. It should not be very frequent, with one set of readings every 10 minutes or so, depending on the type of sensor you use. You can increase this frequency, but there is little reason to bother with these measurements as they will decrease your device's battery life. Additionally, fair use regulations in most countries do not allow continuous streaming in the unlicensed spectrum. Hence, a larger interval is acceptable for this use case, as are most environmental measurements.

The following message is an excellent example of a sample message and the format that it will generally take. The message itself is divided into several parts. The first part contains the ID of the device sending the message. This `WirelessDeviceId` is created by AWS when you add the device. The message also includes the payload, which is the most important. These are the actual sensor data values that we are interested in. To compact the message and save time in sending the message over the air, the payload is sent as a base64-encoded hexadecimal string in a format defined by the sensor vendor. Usually, this string contains information such as the device's battery level and measurement values. It may also include additional information, such as the message type or something specific to the sensor itself:

```
{
    "WirelessDeviceId": "49d70523-11f3-49dd-951c-c90daa11b831",
    "PayloadData":
 "A2fXAARoVAUAAQbLBQd9AgQIfTsACXP8Jgp9CAALfQ4ADH0OAA==",
```

Next is more metadata for the message. This information is pertinent to the protocol and message transfer itself. It provides information to inform you how fast the message was transmitted and some of the parameters about the potential range of the sensor, `Bandwidth`, and `DataRate` of the message. Our interest is in `DevEui`, the device's EUI, usually defined by the manufacturer. We will most likely use this to identify which device the data is coming from when we store and search out data.

The second parameter we are interested in for this part is the `Fport` value. The manufacturer also defines this to tell you what type of data this message might contain. Is it a configuration message, or does it have additional data? With many simple sensors, this can be ignored, but in some cases, but may determine your decoding approach for the payload and which messages you might ignore:

```
"WirelessMetadata": {
  "LoRaWAN": {
    "ADR": true,
    "Bandwidth": 125,
    "ClassB": false,
    "CodeRate": "4/5",
    "DataRate": "5",
    "DevAddr": "00c5c819",
    „DevEui": „24e124710b423686",
    „FCnt": 2230,
    „FOptLen": 0,
    „FPort": 85,
    "Frequency": "867500000",
```

For the next part, we have the gateways, which we saw previously. This part will contain a list of gateways that have received this message. Since we only have one gateway set up, there is only one in the list; however, generally, all gateways report to the network server about an incoming message they receive. The network server (cloud) determines which gateway should reply and continue the interaction. This is great information about how a sensor is interacting with all of your gateways and gives you some context on the range and signal strength of your network:

```
"Gateways": [
  {
    "GatewayEui": "24e124fffef239b8",
    "Rssi": -46,
    "Snr": 11
  }
],
```

The final snippet of the message provides information from the network server itself and the type of message decoding that was provided. Something important here, for the moment, is the Timestamp property of the message. We can use this to determine when the system received the actual message. Another parameter in this snippet is SpreadingFactor. A lower number for SpreadingFactor allows you to send messages faster but can make it harder for the gateway to receive the message due to increased noise. Both the SpreadingFactor and Bandwidth (kHz) properties that were noted earlier, when combined, affect the rate at which data is sent or DataRate. An increase in data rate means less time in the air, but we also need to take battery life and range of the sensors into consideration in these settings:

```
"MIC": "232f13f9",
"MType": "UnconfirmedDataUp",
```

```
        "Major":  "LoRaWANR1",
        "Modulation":  "LORA",
        "PolarizationInversion": false,
        "SpreadingFactor": 7,
        "Timestamp": "2022-03-12T12:25:23Z"
      }
    }
  }
```

This is a simple example of a data upload; a lot of information is wrapped up in this little message. Many parameters are adjustable, and as you expand your knowledge, you will probably customize the parameters within your environment. It is nice to know that things more or less just work out of the box and that customization can come later, giving you time to learn.

As noted previously, we are just interested in a handful of these parameters: PayloadData, DevEui, and Timestamp. This information lets us know where, when, and what information is being sent to us. We also need to know the device type so we can decode the payload appropriately. Some sensors use other parameters, such as FPort to determine the kind of message, but that is not necessary in this case.

We now have a complete LoRaWAN network running. I expect that you are ready to configure additional sensors, either of the same type or different types, to compare and evaluate your options. Next, we need to make use of the data we have collected. Raw message data displayed on an MQTT flow is a great start and shows that you have things working correctly, but for now, we need to decode, store, and visualize that data for analysis.

Collecting data in the cloud

Now that we have data flowing into our AWS account, we can process and store that data practically. *Figure 4.11* illustrates the process and services we will use for this part of the architecture:

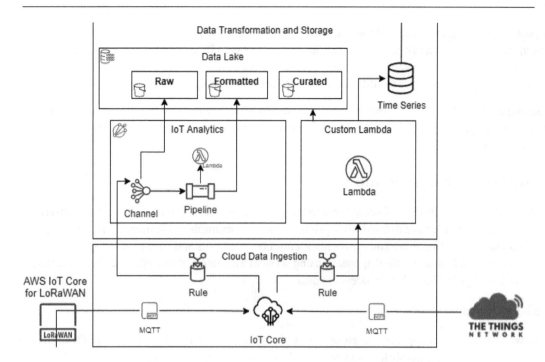

Figure 4.11 – Capturing data in AWS

We will look at two simple ways to start collecting and processing data, putting together the initial stages of a data lake and storing data persistently and efficiently in a simple time series database for quick and easy visualization.

Cloud data ingestion

As you have seen, getting data using LoRaWAN is not very difficult. Once the gateways and sensors are created and configured, the data will flow automatically. Remember that we took two paths to show that there are alternatives available for this option. One option was using AWS IoT Core for LoRaWAN directly to send data. This option works out of the box once things have been set up.

Another option was to configure and use TTN as our LoRaWAN network. This option requires a little more setup that we didn't completely follow end-to-end. We configured a gateway; from there, we can set up applications and add sensors. Data will flow from the sensor to TTN pretty much automatically. Getting that data to AWS requires some extra work.

The integration between TTN and AWS IoT seems to be getting a lot better with time. The integration is tighter and cleaner, and somewhat easier to manage. In some ways, there is additional capability beyond just using native AWS IoT Core for LoRaWAN. TTN has been around for some time and focuses solely on LoRaWAN, so the feature set is a little more robust for the time being. To get more information on how to perform the integration and some of the options, TTN provides constant updates to its integration pages at the following website: https://www.thethingsnetwork.org/docs/applications/aws/. This integration can get a little complex since you need to run AWS CloudFormation templates to get everything set up. We were impressed by the level of integration and the ability to monitor all the different aspects of communicating with the sensor.

Getting data into the cloud for processing is a good step. However, the data is still not very useful to use as-is. We need to process, transform, and store the messages so that they can become helpful as our system evolves.

Data transformation and storage layer

Picking up data from AWS IoT Core for processing allows us to process and store the data as it streams into the platform. We have defined two options within this example to illustrate what is available. Both options, including these and others not mentioned, have pros and cons for each approach. In this case, we will look at a typical approach using services available within AWS, and then a custom approach using AWS lambda to process the data.

IoT Analytics

We will touch on this approach as an option in this chapter. In *Chapter 8, Asset and Condition Monitoring*, we will dive deeply into how to use IoT Analytics fully to process incoming data. AWS IoT Analytics is a managed AWS service that allows you to process, store, and analyze IoT data primarily out of the box. *Figure 4.11* shows two phases in the IoT Analytics service and provides a simplified view of how the service works and how it can benefit your architecture. IoT Analytics works using the following components:

- **Channel**: A channel is the first step in IoT Analytics and allows the service to ingest raw message data from different sources to start processing the data. These messages are dumped automatically into a raw data lake setup in S3 to ensure secure message recording and storage.

    Data is sent to a channel from IoT Core using a rule. Rules can be defined to send messages to an IoT Analytics channel and several other AWS sources for processing and storage. Generally, you may use a different channel for different types of messages or various sensors. This will allow you to store and forward messages for processing based on the format and type.

    Generally, not a lot of data processing occurs within the channel; it is used as a starting point within IoT Analytics to store raw data and get the data into a pipeline for processing. This is a bit of a shame since some data may need to be stored before initial saving or may not be in the format you need for raw storage. But raw is raw, regardless of whether it is what you want.

- **Pipelines** are where all the real work is done in processing IoT data. Channels forward data to a pipeline after storing the raw message in S3. A pipeline can perform several actions, including transforming the data, usually by using a custom lambda function, keeping the transformed data into a new data lake as formatted or transformed data, and then performing additional steps for processing or storing data into a time series data store for analysis or display.

Pipelines are an easy way to get your data through your application's transformation and processing phases. Like most things in AWS, with this great power comes great responsibility. This means that items like decoding or enhancing your data are generally a custom effort. This is something that you have to do since only you know the following things:

- How the data needs to be transformed – that is, what the message decoding sequence should be

- How do you want to enhance, store, and use your data and measurements

None of this should be a surprise. This is how AWS generally works – by providing robust services that allow you to build a customized architecture based on need. The services provided within AWS are constantly evolving with new features and capabilities. But you need to know how they work and how you want to use them to build an effective architecture.

> **Important note**
> When things do not seem to work within AWS, the problem is almost always permission-related. Services within AWS are nicely isolated, and service integration, such as the ability to invoke a function or service from another, needs to be explicitly defined within a policy. CloudWatch is an excellent place to start if you think some permission is being denied in your process.

It's all part of the learning process, so keep a good attitude, even though sometimes, when we are in a hurry, it can be frustrating. One tip would be to keep a log of your naming conventions for roles and policies in IAM; all your service instances. As your environment grows, things can quickly get out of hand, and finding or understanding how things are connected can be tedious. Spending time upfront to define a convention on how things are named can save everyone time in the long term. Your colleagues will appreciate the effort. And anyway, isn't that what architects do?

Custom lambda

IoT Analytics can be a great way to manage and control data flow within and between various systems for complex pipelines. However, there is still the need to decode your messages and transform the data if necessary. We use custom lambda functions to process the messages and perform all the pipeline functions simultaneously. It's an option and allows full customization of the data.

It is slightly more work, but a more significant drawback may be that making changes could be more difficult. If you want to adjust the pipeline, it could be much easier to do it in IoT Analytics than within a bunch of custom code. It is a trade-off to consider when testing your options. You will see later that IoT analytics often requires custom Lambda functions, so the trade-offs are not that drastic.

> **Important note**
> The full lambda code will be downloadable with the purchase of this book; however, I will qualify in advance that this is not necessarily production-quality code. This particular sample code was written as a rough example to show how you might accomplish the tasks in a streamlined manner. I expect your code to be of much higher quality.

We walk through some of the significant sections of the code to give you a feel for the process. The core lambda function for processing the message is written in Python and has several areas working together to do the following:

- Store the original record in the raw data lake
- Decode the payload of the record
- Store the decoded and stripped-down message in the formatted data lake
- Store the decoded message values in a **Timestream** database for visualization

These steps can be recreated later using IoT Analytics, but for now, it is nice to see how this can be manually developed. Calling the initial lambda function can be done by an IoT rule that points to the specific lambda function you have designed; when a message arrives with the correct MQTT topic, the rule fires and forwards the message to your lambda function.

The first part of the function is pretty simple – we look at the incoming message and begin to strip out key values that we need for formatting and processing:

```
def lambda_handler(event, context):
    print("Received event: " + json.dumps(event, indent=2))
```

Remember that we need `Payload`, `DevEui`, and `Timestamp`. `DevEui` and `Timestamp` are used for multiple purposes. Later in the process, they are stored in AWS Timestream as dimension data to help you query your measurement values. These values are also used when storing the data in S3 as partitions to identify the folder structure for message storage to make it easier to analyze and retrieve data stored in the data lake.

The following section of code extracts the set of values used to set key partitions (S3 folders) for storing our data in a meaningful way. When we store the data in S3, we use a folder structure similar to `/<deveui>/<year>/<month>/<day>/`:

```
    #parse out initial key values
    payload64 = event.get("PayloadData", None)
    deveui = event["WirelessMetadata"]["LoRaWAN"].get("DevEui",
None)
    timestamp = event["WirelessMetadata"]["LoRaWAN"].
get("Timestamp", None)
    #strip out timestamp values (probably a more efficient way
of doing this)
    format =  „%Y-%m-%dT%H:%M:%SZ"
    dt_object = datetime.datetime.strptime(timestamp, format)
    year =  datetime.datetime.strptime(timestamp, format).year
```

```
    month =  datetime.datetime.strptime(timestamp, format).
month
    day =  datetime.datetime.strptime(timestamp, format).day
    minute = datetime.datetime.strptime(timestamp, format).
minute
    hour =  datetime.datetime.strptime(timestamp, format).hour
    second = datetime.datetime.strptime(timestamp, format).
second
```

One note on the partition structure, this simplified structure doesn't really follow a unified namespace convention. Ideally, you would need a more robust partition structure to accommodate an enterprise's worth of data.

I admit parsing out the date/time information was a little awkward. Sometimes, programming is just tedious, so we plow through it step by step. Next, the raw message is dumped into the initial S3 data lake as raw storage. This takes the initial message and stores it in the `s3-datalake-iot-raw` bucket. Notice that we use the year, month, and day information from the previous section of the code to store our data. Ideally, this message is now stored in the `lorawan/am319/<deveui>/<year>/<month>/<day>/<hour>/` folder structure with a filename of `<minute>:<second>.json`:

```
    #dump initial message into raw datalake
    s3 = boto3.client("s3")
    bucket_name = "s3-datalake-iot-raw"
    file_name = str(minute) + ":" + str(second) + ".json"
    folder_path = "lorawan/am319/deveui=" + str(deveui) +
"/year=" + str(year) + "/month=" + str(month) + "/day=" +
str(day) + "/hour=" + str(hour) + "/" + file_name
    s3.put_object(Bucket=bucket_name, Key=folder_path,
Body=json.dumps(event, indent=2))
```

Next, we need to transform the payload into a readable format. We will talk more about the actual decoding process later in this chapter, but for now, we must call a second lambda function that is designed solely to decode this message format. We use this second decoder message because it is easy. Often we need to figure out the decoding ourselves, and we will discuss that in a later chapter, but in this case, the vendor provided this decoder:

```
    #pass the event message to the decoding lambda
    response = lambda_client.invoke(
        FunctionName = 'arn:aws:lambda:us-west-
2:xxxxxxxx:function:iot_decode_lorawan_raw_am319',
        InvocationType = 'RequestResponse',
```

```
        Payload = json.dumps(event)
    )

    decode_response = json.load(response['Payload'])
        response_json = literal_eval(json.dumps(decode_
response, indent=2))
    message_body = response_json.get("body")
```

The decoded `message_body` variable holds the response from the second lambda function. This is our decoded set of values derived from the payload of the sensor. The `decoder` function also added the two values of `deveui` and `timestamp` that we want to include in our final message data:

```
{"temperature":21.5,
"humidity":48.5,
"pir":"idle",
"light_level":3,
"co2":1084,
"tvoc":34,
"pressure":1009.9,
"hcho":0.09,
"pm2_5":22,
"pm10":24,
"deveui":"24e124710b423686",
"timestamp":"2022-02-27T13:14:31Z"}
```

Once we have decoded the message payload, we can store the data in the data lake S3 bucket for formatted data. Of course, we could continue our effort and enhance the data with more information from additional sources:

```
    #store transformed message into formatted data lake
    bucket_name = "s3-datalake-iot-formatted"
    file_name = str(minute) + ":" + str(second) + ".json"
    folder_path = "lorawan/am319/deveui=" + str(deveui) +
"/year=" + str(year) + "/month=" + str(month) + "/day=" +
str(day) + "/hour=" + str(hour) + "/" + file_name
    s3.put_object(Bucket=bucket_name, Key=folder_path,
Body=json.dumps(message_body, indent=2))
```

Finally, we can write the measurement data to a Timestream database to get some value from the data almost immediately. We extract some values to set up dimensions for our inserts into Timestream. Dimensions are metadata attributes for the time series data. In our case, the source and type of sensor provide the data. In addition, we add DevEui to query the data values easily. There are many more dimensions we will need to add as we continue, but this is enough to get started:

```
#Store data into timestream
timestream_values = json.loads(message_body)
#picking up current time, since Timestream does not allow
for prior dates. Just a one-off since I am reusing old data as
message input.
#as long as there is not too big a gap this should not be a
problem.
current_time = str(int(round(time.time() * 1000)))
deveui = timestream_values.get("deveui", None).strip('"')
original_timestamp = timestream_values.get("timestamp",
None)

#Dimensions are required for every timestream entry.
dimensions = [
    {'Name': 'device', 'Value': 'am319'},
    {'Name': 'source', 'Value': 'lorawan'},
    {'Name': 'deveui', 'Value': deveui}
]
```

Formatting the data so that it can be inserted into Timestream is a simple enough set of tasks. We define each measurement and corresponding information around the name/value pair of the measurement. In this part of this example, we have provided the full structure for the temperature measurement; however, we have constrained the additional measurements as placeholders only. Notice that the Dimensions array is added to every record as a standard across all measurements:

```
temperature = {
    "Dimensions":dimensions,
    'MeasureName': 'temperature',
    'MeasureValue': str(timestream_values.
get("temperature", None)),
    #'MeasureValue': '21.5',
    'MeasureValueType': 'DOUBLE',
    'Time': current_time
}
```

There is a similar section for each measurement, as shown here. Please download the code sample for this chapter to see the complete example:

```python
humidity = {…}
pir = {…}
light_level = {…}
co2 = {…}
tvoc = {…}
pressure = {…}
hcho = {…}
pm2_5 = {…}
pm10 = {…}
```

Once the measurement records have been created, we can write our data to the Timestream database:

```python
records = [temperature, humidity, pir, light_level, co2,
tvoc, pressure, hcho, pm2_5, pm10]
try:
    result = timestream_client.write_
records(DatabaseName='am319', TableName='data',
Records=records, CommonAttributes={})
    print("WriteRecords Status: [%s]" %
result['ResponseMetadata']['HTTPStatusCode'])
except timestream_client.exceptions.
RejectedRecordsException as err:
        print("RejectedRecords: ", err)
        for rr in err.response["RejectedRecords"]:
            print("Rejected Index " +
str(rr["RecordIndex"]) + ": " + rr["Reason"])
        print("Other records were written successfully. ")
except Exception as err:
        print("Error:", err)
```

That's it. This example is simple and functional, without overloading you with a lot of additional processes. We skipped through some extra stuff that is needed at the beginning and end of the function, but please download the code and review those sections.

> ## What about Node-RED for data processing?
>
> I was going to include a section on using Node-RED to illustrate using it to process TTN data, but in researching this chapter, it seemed unnecessary since the latest integration between TTN and AWS IoT Core looks pretty solid. While it is not needed in this context, I thought it might be helpful to say a few words from experience.
>
> I sometimes kick myself (not very hard, though) for not recognizing the power of Node-RED earlier. I was introduced to the technology early on and dismissed it pretty quickly as not being scalable and seemingly something that would not catch on widely. In my defense, I was looking at Node-RED very early in its development and initial release, and, like most of us, I had a lot going on at the time. Adopting a new, unproven technology seemed unwarranted. I wasn't a great JavaScript programmer (I'm still not!), and I couldn't get guidance on how it would scale horizontally if and when I needed it. Experienced architects always have this dilemma – that is, determining which horse to back as they consider a product or solution.
>
> Several years (actually, many years) later, I came across a problem where Node-RED was presenting itself as the solution. I needed to process many data streams from a LoRaWAN network with a standard and cost-effective approach. After some work setting things up, Node-RED came to my rescue, and I was a believer. Launching one instance of a Node-RED server allowed me to process many streams of data quite efficiently by listening to various topics, processing, and then forwarding each message to its destination.
>
> This wasn't a high-volume situation but rather something I wanted to be cost-effective, so I wasn't worried about scale, but the flexibility and ease of setting this up and the initial decoding and processing of the messages wowed me considerably. Live and learn!

There is a lot of detail here about the overall process that we simply cannot cover, especially when it comes to using general AWS services such as lambda – for example, how to create a lambda function. This is generally available knowledge that can be looked up and may change over time. It would be redundant to share that here as this is not designed to be a step-by-step tutorial.

We skimmed over decoding the payload in the previous section, but this could be considered the most important part. Let's spend a few minutes now and see how we can approach the decoding effort.

Decoding the payload

Payload decoding is a mystic art, as are many things in computer science. Most payloads are base64 hexadecimal strings that must be decoded and translated byte by byte. Each sensor is different because the vendor defines the measurements, values, and order. For example, consider the following hexadecimal string:

```
"PayloadData":
"A2fXAARoVAUAAQbLBQd9AgQIfTsACXP8Jgp9CAALfQ4ADH0OAA==",
```

The goal is to turn this into ASCII and formatted in JSON:

```
{"temperature":21.5,"humidity":48.5,"pir":"idle","light_level"
:3,"co2":1084,"tvoc":34,"pressure":1009.9,"hcho":0.09,"pm2_5":
22,"pm10":24,"deveui":"24e124710b423686","timestamp":"2022-02-
27T13:14:31Z"}
```

The latter string represents the information we most need from the message: the payload data itself, the `deveui` property or ID of the sensor, and the date and time of the message. Fortunately, most sensor developers provide guidance for decoding messages and interacting with the sensor. Milesight is no exception, providing a detailed interactive guide at `https://resource.milesight-iot.com/milesight/document/quickguide/am300-user-guide-en.pdf`.

Embedded within this guide is a table similar to the following, which provides a detailed breakdown of each byte and how it can be interpreted into a measurement value:

Item	Channel	Type	Description
Temperature	03	67	INT16, Unit: °C, Resolution: 0.1 °C
Humidity	04	68	UINT8, Unit: %, Resolution: 0.5 %
PIR Status	05	00	01: PIR is triggered 00: PIR is not triggered
Light Level	06	cb	00:0-5 lux 01:6-50 lux 02:51-100 lux 03:101-500 lux 04:501-2000 lux 05:≥ 2000 lux
CO2	07	7d	UINT16, unit: ppm
TVOC	08	7d	UINT16
Barometric Pressure	09	73	UINT16, unit: hPa, Resolution: 0.1 hPa
HCHO	0a	7d	UINT16, Unit: mg/m3 , Resolution: 0.01 mg/m3
PM 2.5	0b	7d	UINT16, unit: μg/m3
PM 10	0c	7d	UINT16, unit: μg/m3
O3	0d	7d	UINT16, unit: ppm

Table 4.1 – AM319 message payload format

By decoding the original base64 payload string, we get a string of hex digits similar to this:

```
0367D70004686105000006CB03077D3C04087D2200097373270A7D09000B7
D16000C7D1800
```

This string maps directly to the preceding table. The first 4 bytes consist of the channel, the type, and the value, D7 = 215, in decimal. Divide by 10, and you get the temperature in Celsius at 21.5 degrees. Again, *decoding messages is a mystic art*. There is no doubt about that. For this particular sensor, the sample code for fully decoding the payload values is available online at `https://github.com/Milesight-IoT/SensorDecoders`, where Milesight has provided several decoders for use with their sensors.

Instead of reinventing the wheel and writing decoders for this example, we used one of the downloadable examples and dropped it into a separate lambda. This was expedient since the function did the more complex work of decoding the payload, and we ended up with two lambda functions for this sensor with a good division of responsibilities. They are as follows:

- `Iot_decode_lorawan_raw_am319`: This function decodes the actual payload and reformats the response without all the extra data we don't need downstream in data analysis.

- `Iot_process_lorawan_am319`: This function processes the whole message end-to-end, storing and decoding data along the way. This function is similar to what we would achieve with an IoT Analytics pipeline.

By splitting some of the functionality, we have already started to design an approach based on a separation of concerns. Different functions can be used in different ways. For example, the `iot_decode_lorawan_raw_am319` function can be used by the IoT Analytics pipeline, the rules, or our lambda function.

We have done some basic processing and stored the data as it was collected. But now, we are going to see the data in a meaningful way. In the next section, we will spend some time setting up a simple dashboard to visualize our effort immediately.

Setting up a basic dashboard

For initial visualization, there are several options within AWS. One good option for a long-term **Business Intelligence** (**BI**) service is **AWS QuickSight**. QuickSight allows you to connect to your data and includes in-memory data storage for quick data analysis or to provide customized dashboards and reports. No proper BI tool (AWS or otherwise) is ready out of the box for end users. Transforming and loading data sources and end-user training to make good use of the data and build valuable visualizations is an effort.

We will explore AWS QuickSight in a later chapter, so for our initial setup, we will build a simple dashboard to show the incoming data and to get a feel for how the information looks. We will do this by starting simple and then combining data in ways that make sense—energy versus flow, or machine parameters that can be compared against quality, can be a valid combination of measurements. *Figure 4.12* shows a simple AWS **Grafana** dashboard of our deployed sensor. Remember that the AM319 provides nine different measurements at regular 10-minute intervals. AWS Grafana integrates easily with the Timestream database we created to store data and allows you to set up a simple dashboard showing the latest measurement values. Using Grafana to view time series data is an excellent fit for the technology for our immediate purpose:

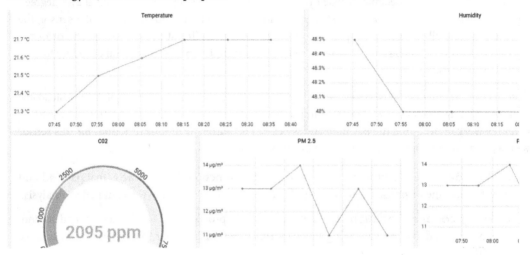

Figure 4.12 – Grafana dashboard

In our case, this display consists of the temperature, humidity, CO_2 levels, PM 2.5, and PM 10. Only a small set of measurements is displayed on this dashboard. However, it is straightforward to set up and display on a monitor in an office or plant. Grafana also makes it relatively easy to build several different dashboards and put them into a rotating playlist, which allows you to group relevant items into a display.

There is also a presence sensor in the device to consider, which detects the infrared radiation emanating from people in the area around the sensor. More commonly known as a **Passive Infrared Sensor** (**PIR**), we could use this to help determine, for example, if the presence of people in an area raises temperature or CO_2 values.

AWS Managed Grafana allows you to connect quite easily to several AWS services, including **CloudWatch**, **IoT SiteWise**, Timestream (our use case), **Athena**, and others. Generally, Grafana is very useful for building dashboards of logs, metrics, or any time-based data, which makes it perfect for this example. However, it may not be the right long-term strategy where you want to combine our IoT data with external data and perform deeper analysis and intelligence. However, the combination of Timestream and Grafana may be a good dashboard approach to viewing real-time data coming from your environment that may expire after a short period. If your Timestream data is short-lived and you are viewing data from the current day, week, or even the last few hours, this part of the solution will be a great fit.

Putting the layers together

Putting all the pieces of the architecture together, you can see that the complete flow from the sensor to the screen is relatively simple. That is our initial goal – getting some data flowing and obtaining a better understanding of what type of environment we are working with. This is just one path for one set of use cases, but it allows us to start formalizing the layers we have defined, to understand or test what services and components we may use, and to vet some of the data needs from our initial environment. *Figure 4.13* provides a view of the complete architecture as we have implemented it so far:

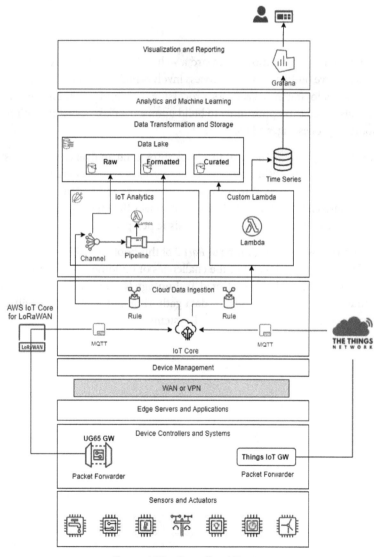

Figure 4.13 – Overall architecture

You can see that some of the layers have been skipped for this example. Our focus is to crawl and walk first and then run later, especially when we are just getting the initial phases of the architecture designed and built. It has been said that architecture is as much of an art as a science, which is as accurate in the IT world as in the physical world. All the layers are outlined in *Figure 4.13*; however, they are incomplete. We have a very narrow stream of data flowing from our sensors, and, at that, the velocity is low, with messages coming in at 10-minute intervals or more. This is fine as a goal for now. We want to get some preliminary data flowing to combine some of the architecture without worrying too much about being overwhelmed with data or thinking too much about analysis.

Summary

There are a lot of moving pieces, but moving toward a well-defined architecture can help you make sense of it. In this chapter, we outlined how the process involves building something from nothing and laying out the steps for deploying initial sensors, processing data, and providing basic visualization to our end users. We also used this opportunity to build upon a data lake structure within AWS as we added new data sources, processing, and features.

With that, we have also wrapped up *Part 1* of this book, providing an overview of Industry 4.0 and the digitization of industry. This exciting new area aims to give the next generation of value through better management and optimization of industrial environments and systems. Our focus in *Part 1* was on the basics of industrial IoT, with a lot of emphasis on *why* digital integration can be significant and *how* we might approach our goals from a more holistic perspective.

In the next chapter, which is also the beginning of *Part 2* of this book, we will continue to dive deeper into industrial systems and address many of the challenges of building systems and managing IoT data at scale. We will continue into the **Operational Technology** (**OT**) realm and consider integration patterns with standard industrial protocols for data gathering and analysis. We will also explore different integration approaches and designs for IoT systems.

Part 2:
IoT Integration for Industrial Protocols and Systems

In the previous section, we focused on the basics of IIoT, emphasizing why digital integration is essential and how we might approach a project from a holistic perspective. In this section, we will explore the operational technology realm and consider integration patterns, with standard industrial protocols for data gathering and analysis. Along the way, we will explore different industrial integration approaches and some options for integration into an IoT system.

This section focuses on understanding and connecting to industrial devices and learning more about how they work, and the value derived from interacting at the machine level. Some basic knowledge about PLCs, how they work, and how to interact with them will help IT understand the challenges of the OT environment. We provide real-world examples of defining, capturing, and processing industrial data to help you determine your solution architecture.

In addition, we will dive back into our hands-on examples in the AWS cloud, and we will show how to gather simulated data and process it for operators to display and use to understand their environment better.

This part of the book comprises the following chapters:

- *Chapter 5, OT and Industrial Control Systems*
- *Chapter 6, Enabling Industrial IoT*
- *Chapter 7, PLC Data Acquisition and Analysis*
- *Chapter 8, Asset and Condition Monitoring*

5

OT and Industrial Control Systems

OT stands for **Operational Technology**, used in industrial production. This chapter will focus on existing control systems prevalent in the industry, commonly attributed to the OT layer. These are part of layers 0, 1, and 2 of the Purdue model and ISA-95 architecture. This chapter will help you understand real-time manufacturing execution systems and how machines are orchestrated to maximize production.

In *Chapter 2, Anatomy of an IoT Architecture*, we outlined a layered approach to architecture. This chapter will focus on the lower layer and move toward an end-to-end integration. The device layer, or the lowermost layer, consists of machines that operate, PLCs that help the machines run, and the SCADA and control systems that supervise the PLCs and equipment.

We'll start with understanding the landscape of a typical manufacturing unit with machinery and control systems and then perform a deeper dive into the intelligent controllers that drive the machines to achieve production goals. We will also explore the language of these machines – the underlying protocols that define the all-important communication mechanism.

In a nutshell, we will look at the following areas:

- Revisiting the basics of manufacturing
- Understanding industrial control systems
- Exploring PLCs
- PLCs versus microcontrollers
- Industrial communication of the future
- Mapping your integration strategy
- Where IT meets OT in industrial systems

Technical requirements

This chapter provides an overview of industrial control systems. First, we deconstruct a PLC to understand the inner workings of the automation device that single-handedly fueled Industry 3.0. The second half of the chapter is dedicated to advanced topics that drive the industry today and will drive the industry tomorrow. A basic understanding of industrial networking would be a plus for the second part of the chapter. Knowledge of the following would be a great starting point to better understand this material:

- Control systems
- Fieldbus protocols in the OT domain
- Basic networking concepts
- Security fundamentals

Revisiting the basics of manufacturing

Manufacturing, in general, can be classified as discrete manufacturing or process manufacturing. While discrete manufacturing deals with creating countable or distinct products, such as cars or shoes, process manufacturing usually follows a recipe combining multiple raw materials to create different finished goods. Examples of this include detergent, face cream, or steel. A keen understanding of various manufacturing processes is key to appreciating the importance of different control systems used for each of them. *Figure 5.1* visualizes the multiple subdivisions of the kinds of manufacturing arranged in decreasing order of production volume and increasing order of production variety.

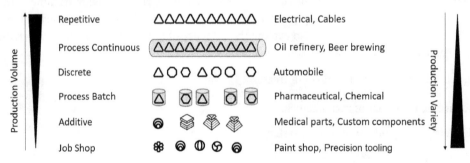

Figure 5.1 – Manufacturing types

- **Repetitive manufacturing**: In repetitive manufacturing, the production line is tuned for manufacturing a single **Stock Keeping Unit** (**SKU**) with no or minimum variations. A one-to-one mapping between the production line and the product itself results in mass manufacturing (fast turnover) and minimal changeover. The disadvantages of this approach and setup are the lack of flexibility with the changing times (demand, relevance, etc.) and the lack of customizability of the product.

- **Process continuous manufacturing**: Oil refineries and breweries come under this manufacturing category, where multiple processes are carried out to produce a single variant of the finished good. It is like repetitive manufacturing in terms of the capability to run at a specific rate, with some differences attributed to the nature of the finished goods. We can still dismantle the finished goods in discrete manufacturing into their constituent sub-parts. By contrast, in process-based manufacturing, the elements get mixed physically and chemically to form a new product that is inseparable from its base components.

- **Discrete manufacturing**: Discrete manufacturing involves the capability to produce many different SKUs on the same production line or within a facility. For example, the same assembly line could assemble a red, diesel, four-airbag variant of a car while it could also assemble a blue, petrol, seven-airbag variant. While adding the ability to customize, this manufacturing process can add some delay due to changeovers of equipment, stations, tools, spare parts, or the procurement of raw materials.

- **Process batch manufacturing**: Batch manufacturing allows for different products getting produced in the same production line. For example, you could manufacture scented hair oil in the morning while making body lotion during the second shift. **Cleaning-in-place** (**CIP**) is a critical, mandatory process in these manufacturing types to avoid contamination and improve quality output.

- **Additive manufacturing**: Advancements in 3D printing technology and the reduction in costs have greatly improved the adoption of this type of manufacturing. Custom 3D-printed soles for your shoes are a great example of additive manufacturing, where user-generated requirements are individually produced. Additive manufacturing also has great value in the prototyping industry to create custom low-quantity batches. Another use case is the manufacturing of spare parts, which can be difficult to ship or import.

- **Job shop**: Job shop dates back to the beginning of manufacturing when every product was hand built. Any product that needs a custom procedure goes into a job shop. Precision tooling, specific customization add-ons, and painting are good examples of this work. While the volume is possibly the lowest with this type of approach, customizability is at its highest.

Since there are different methodologies involved in manufacturing, the control systems are also complex and computationally intensive. Each type of manufacturing system must be approached differently, as there cannot be a one-size-fits-all solution. Luckily for us, **Programmable Logic Controllers** (**PLCs**) have been historically designed as general-purpose controllers that can be explicitly programmed to work with the equipment and solve the problem. Let's look further into industrial control systems and how they work.

Understanding industrial control systems

Industrial control systems have existed since the third industrial revolution, enabling rapid mass production of goods assisted by computer interfaces. They consist of the hardware and software responsible for real-time operations at the machine level. The PLC, **Remote Terminal Unit** (**RTU**), and **Embedded Controllers** are the first stage of control within the manufacturing environment.

Supervisory Control and Data Acquisition (SCADA) generally sits one layer above to supervise and control a network of PLCs and other devices. **Distributed Control Systems (DCSs)** combine PLC and SCADA in a process-oriented manufacturing setup. For the scope of this book, we will be doing a deep dive into PLCs while giving a limited overview of the other high-level manufacturing systems, except to discuss where we might gather data from some of these systems.

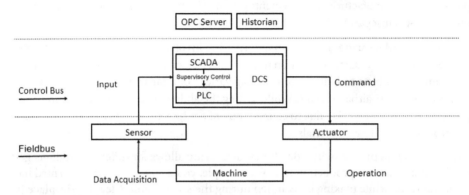

Figure 5.2 – Industrial control systems overview

PLCs, SCADA systems, and DCS are part of levels 0, 1, and 2 of the ISA-95 architecture model, as presented in *Figure 5.2*.

PLCs are small industrial computers that sense data from input devices and actuate output devices based on built-in logic. They are at the bottom of the information loop and are the building blocks of the data fabric. They are very robust, exceptionally reliable, and operate in near real time, powered by a proprietary operating system.

PLCs do not have a structured, comprehensive data model, resulting in increased integration difficulty. Data elements, variables, or device addresses within a PLC are assigned names or tags. These PLC tags do not generally have a standardized nomenclature, which puts us at the mercy of the installation team and independent system integrators in detangling a spaghetti of tag names. This can make connecting and understanding data from multiple PLCs strenuous.

Due to the nature of most production facilities, there might be PLCs from multiple **Original Equipment Manufacturers (OEMs)** and PLCs from the same OEMs across multiple generations. This proves to be a highly heterogeneous and complex field in which to play. In the industry, this is known as a **brownfield environment**, which may have a long history of systems and components that are outdated and incompatible.

Some PLCs might also have locked program ports due to multiple **System Integrators (SIs)**, rendering them practically useless from a data acquisition standpoint. PLCs might also be a potential attack surface during a cyber threat and can result in more significant harm if not secured. Well-planned and architected security paradigms are needed for a successful and secure interconnection. Another major challenge is the multitude of protocols (Ethernet/IP, MODBUS, PROFINET, PROFIBUS, and so on) prevalent for a PLC, especially in a brownfield environment. Thus standardization is essential and can provide more direction for the way forward.

SCADA is one level above the PLC network, which connects and controls a PLC network and its activity. SCADA systems are generally mapped to PLC tags and follow a standard data model, with parameters such as tag name, description, and sampling time set-points to handle thousands of parameters. They also have multiple integration points for **Human/Machine Integration** (HMI) and for talking to higher-level systems, such as **Manufacturing Execution Systems** (MESs) and **Enterprise Resource Planning** (ERP).

SCADA systems are well-tuned and run perfectly as closed-loop systems. However, there can be drawbacks. They historically run on Windows machines, which can have real-time performance and availability issues. Real-time performance cannot always be guaranteed due to operating systems' timing issues, reboots, and patch updates. SCADA systems are also built with several modules that execute different tasks, which might get overloaded and unresponsive at scale. Since these are often closed-loop systems, connector development, maintenance, and backward and forward compatibility can all be issues that plague our effort.

DCS is a combination of PLCs and SCADA systems in one. They are predominantly used in process-based industries such as refineries. DCS has a hierarchical data management system to simplify data management, making them more efficient than PLCs. In terms of downsides, DCS suffer similar issues ranging from multi-vendor and multi-generation ecosystems to connectivity issues.

With this introduction and analysis of the overall environment, let us look deeper into PLCs to understand them better in the context of industrial automation and Industry 4.0 applications.

Exploring PLCs

PLCs are the fundamental hardware components in an industrial control system. These purpose-built controllers kick-started what is collectively termed the Industry 3.0 revolution. They were single-handedly responsible for automating the shopfloor and multiplying production capabilities, thereby disrupting the global economy.

Starting with the history and evolution of PLCs, we will deconstruct a PLC to its components and look at software control in this dedicated section. We will also present an interesting comparison between PLCs and the latest DIY **Single-Board Computers** (SBCs) in vogue among the IoT maker community.

History and evolution of PLCs

The evolution of PLCs started in the mid-1950s, shown in *Figure 5.3*. The relays that the PLCs replaced were complex, bulky, and expensive to maintain and control. Dick Morley, regarded as the father of the PLC, created the first prototype called the standard machine controller in 1968. Parallel developments happened, and Allen Bradley introduced the 1774 family of PLCs in 1971. PLCs evolved further to have better CPUs, I/O expansion cards, and rack mountability.

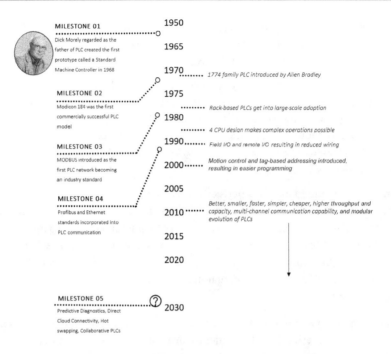

Figure 5.3 – Evolution and milestones of the PLC

Introducing Modbus as the de facto industry standard was a significant milestone, creating a path for interoperability and data homogenization. Modbus was designed as a communication protocol for data exchange between PLCs using a master/slave approach. Data was initially transferred using a binary stream over a serial connection between devices but has evolved to using Ethernet with most systems today. As part of the unification and standardization initiative, the **International Electrotechnical Commission (IEC)** 61131, an IEC standard for programmable controllers, was introduced. It was first published in 1993, with the most recent (third) edition published in 2013.

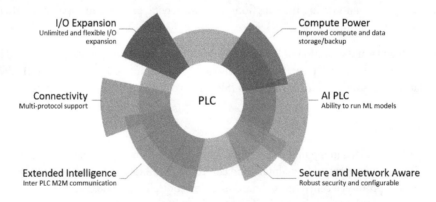

Figure 5.4 – PLC evolution areas

The PLC OEM industry is very competitive, and much effort is spent in R&D to create the next family of breakthrough PLCs. While some may consider that the evolution of PLCs has plateaued, we feel it's time for the next leap forward: PLCs with built-in intelligence, predictive diagnostics, and a direct secure connection to the cloud. PLCs that talk to each other within the network and are capable of collaborative, swarm-based decision-making in real time are the next steps in the evolution and future predictions of the exciting times ahead. This is shown as a quick graphic in *Figure 5.4*.

Soft PLCs

Traditional PLCs have often been constrained by limited memory and compute power. Soft PLCs are a segway to harness the theoretically unlimited storage and compute power of a PC or high-end embedded hardware. Soft PLCs can take advantage of multiple Ethernet ports, RAM, and enhanced storage to execute real-time firmware for machine control. Traditional ladder logic programs are run by the Soft PLC core, making them compatible with existing code. The single major advantage of a Soft PLC is that it can reduce the cost of special-purpose hardware I/O modules while still being capable of the deterministic performance of a traditional programmable logic controller. Upgrading from legacy systems, process optimization, and custom implementations are all benefits of Soft PLC technology.

Deconstructing PLCs

As defined earlier, PLCs are miniature industrial computers capable of performing control automation operations in real time through an amalgamation of software and hardware. The **National Electrical Manufacturers Association** (**NEMA**) describes a PLC *"as a digital electronic device that uses a programmable memory to store instructions to implement specific functions such as logic, sequence, timing, counting, and arithmetic operations to control machines and processes."*

Controlling machinery at a production facility, running continuous assembly lines, and powering brewing plants are various applications of the PLC. They are designed to read environmental conditions based on digital or analog inputs and act based on set points, making them intelligent condition-based operators. They provide an interface and a control point between input sensors and output devices actuating based on logical decisions set up in their programming.

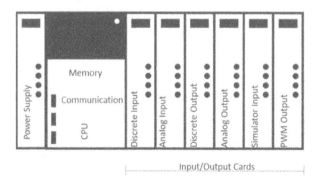

Figure 5.5 – PLC: The components

The various components of a PLC are visually represented in *Figure 5.5*. The core of the PLC is the CPU or the processor module. This consists of a microprocessor platform with a **real-time proprietary operating system (RTOS)** embedded in it. A typical desktop operating system has a lot of overhead and continuously works in the background on various activities. It switches its activity, putting some actions on hold while it attends to something else. An RTOS is designed to be clutter-free, focusing solely on real-time data acquisition and actuation that the production facility demands. It doesn't have any other purpose and will not be distracted from the assigned task. The processor module also has internal and external storage capabilities for logging, monitoring, and observability. These are important functionalities to assess the real-time performance and operational health of the system.

We will look at practical examples in *Chapter 7, PLC Data Acquisition and Analysis*. The examples will explore basic PLC programming and application-based complex programs. The processor also has built-in communication modules, such as Industrial Ethernet, Modbus, and RS485, to help move data where it can be examined and processed more effectively. While in programming mode, the PLC accepts a compiled program from a PC that has the necessary logic for execution while in *Run* mode. The PLC also needs a power supply module for functioning. The power supply provides the necessary electrical isolation to the PLC, guarding it against fluctuations in current and voltage.

In addition to the CPU, the design of modern PLCs allows for stackable I/O cards, which are used based on application needs. They provide flexibility and modularity in cost and space and allow for more direct integration into all industrial systems. Various input and output modules are available for easy rack mounting and commissioning. Temperature, pressure, flow rate, level, and a thermocouple are all analog inputs that can be directly interfaced with a device. Typical analog input values range from 4-20 mA to 0-10 V.

Digital or discrete inputs are enabled through push-button switches, limit switches, beam sensors, and inductive switches. At the same time, outputs can drive analog, discrete, and high-speed **Pulse Width Modulation (PWM)** modules. These are valves, lamps, alarms, solenoid coils, motor starter units, fans, relays, and buzzers. HMI modules can also be connected near the machines to give the operator direct visibility or control of the equipment.

Finally, there will be a system bus, which is the backbone for the integration of the PLC and different modules. Depending on how the PLC modules connect, this may be visible or inferred. This backbone is necessary to send data, address, and control operations between your configuration's CPU and I/O modules.

Let us now delve into how we can program a PLC to monitor the environment and make decisions.

PLC programming

The IEC 61131 identifies five standard programming languages as the most common for PLCs. These are represented in *Figure 5.6*. PLCs typically use something called ladder logic programming. Let us look at this to focus our efforts on the most widely used method of PLC programming.

PLC Programming Methodologies

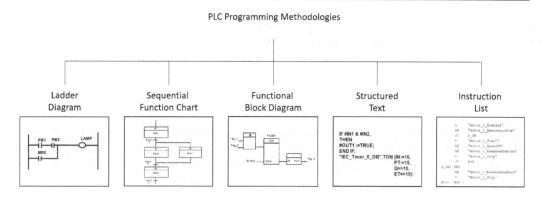

Figure 5.6 – Types of PLC programming languages

Ladder diagrams are one of the most popular and simple of all languages for programming PLCs. They have evolved from the initial relay logic panel and are simple to comprehend and troubleshoot for humans. They resemble a ladder with parallel rungs filled with logic connecting the positive power rail on the left and the negative power rail on the right.

The vertical line on the left is the power, and the right is the ground line. Logical instruction is written in each horizontal line, also known as rungs. Each symbol on the rung is a logical instruction set. A sample ladder diagram is shown in *Figure 5.7*. This is a simple example of switching a motor on and off using two switches: **Start Button** and the **Stop Button**.

As we will see later, the rungs or the horizontal lines are executed sequentially. When the start button switch is pressed, the instruction gets a Boolean positive, and power flows to the next stage. It also checks that the stop button switch is not pressed. Once this condition is true as well, it proceeds to the timer section, where the delay time is executed, after which the motor's output coil is energized. The motor then turns on.

The main constructs of a ladder diagram are the input components, the logic, and the output components through which the power circuit gets completed.

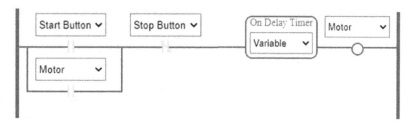

Figure 5.7 – Understanding PLC ladder logic

In *Chapter 7, PLC Data Acquisition and Analysis*, we will look deeper into some practical ladder logic examples with advanced treatments, such as sending alarm emails from a web server running on the PLC.

In the next section, we will look at an exciting topic: comparing PLCs with IoT-enabled low-cost maker hardware. This is a highly debated topic in the current market evolution context. Remember that it is a technology capability comparison and not a product-to-product comparison.

PLCs versus microcontrollers

With recent advancements in the development of low-cost opensource hardware, the DIY or maker community has advanced considerably in an attempt to bridge the OT/IT gap. Tech enthusiasts can now create astounding projects revolving around control systems and sense environment variables; understand, store, and visualize them; and then actuate responses.

The difference, of course, is IoT in the mix: the idea of going beyond the traditional and bringing data to the cloud where it can be examined at a much faster and larger scale. This is similar to how a PLC performs a control operation in a production facility without the cloud. At the same time, machine learning components are needed to bring better insight into the operation. It is logical with this approach to see whether microcontrollers, with their more powerful computational ability, can be substituted. For example, can we exchange a PLC with a general-purpose microcomputer, such as a Raspberry Pi?

In this section, we provide a simple comparison between the two approaches. The choice of hardware is then left to the individual scenario based on the evaluation criteria that form the requirements or constraints of a given application. *Table 5.1* compares and evaluates PLCs and general-purpose SBCs:

Evaluation Criteria	Programmable Logic Controllers	General-Purpose Single-Board Computers
Definition	PLCs are purpose-built controllers specifically designed for automation	SBCs are general-purpose and multi-application oriented
Hardware	Closed hardware with comparatively reduced capabilities (storage, memory, and compute)	Open/closed hardware with higher computational prowess
Operating System	Proprietary RTOS	Open source OS with customizability
I/O Expansion	Simple plug-and-play I/O cards	Usually, custom hardware development is needed
Analog Inputs	Possible through simple I/O cards	Signal conditioning circuit and other data acquisition modules are needed
Reliability	Highly reliable system	Prone to OS malfunctions and CPU overload due to multi-application support

Programming Options	Effective programming through ladder diagrams, h/w, and s/w watchdogs	Multiple options, but not suitable for real time (excluding Python – procedural)
Synchronization	Highly synchronized	Absence of clock synchronization
Northbound Connectivity	Basic in-built. Possible through connectors.	Advanced northbound connectivity
Data Visualization and Control	Possible through HMI cards	Possible in-built – connection to display needed
Industry Readiness	Out-of-the-box plug and play	Not ready for harsh environments for production usage
Cost	High	Generally low cost for the initial board, but more for additional hardware
Verdict	Suggested for industrial usage for mission-critical production activity	Suggested for developing and validating POCs

Table 5.1 – Comparison and evaluation of PLCs and general-purpose SBCs

The preceding table will help you choose the suitable device or strategy for your implementation based on the project stage, budget, time frame, end application, and so on. You might wonder whether the technical advancements in the PLC technology domain have plateaued. We now have an interesting callout for you, talking about the most recent advancements in PLC technology.

Running AI inside a PLC: NPUs

Artificial Neural Networks are computing systems inspired vaguely by biological memory networks of animals. Such systems *learn* to perform tasks by considering examples of events that occur without being pre-programmed with a set of task-specific rules.

Normally, we think of edge computing at the edge of the factory before data moves to the cloud, using more traditional computers. **Neural Processing Units** (**NPUs**) advance the use of AI at the extreme edge, that is, on the PLC! They are powerful compute units capable of running a machine learning model on the PLC, deriving inferences from real-time data. Continuous improvement of the neural network is also possible through retraining the models with the latest process data. This can occur in a global cloud or a local edge. Moreover, through automatic redeployment, the latest models can run on the NPUs.

The application of NPUs is tremendous. Robotic arms can now handle any new and unknown objects through neural networks. AI-based visual inspection agents can learn from their human counterparts – the quality engineers with specific expertise and know-how. This gradual learning can then be incorporated to improve their visual inspection algorithms, making them closer to humans. Detection of process anomalies in near real time is also possible through AI, and predictive maintenance can kick right in.

The next section of this chapter is a core topic that presents a vision for the future of industrial communication. Up until now, you might be aware of the industry's extraordinarily complex and heterogeneous nature with multiple vendors, protocols, consortiums, and working groups. Arriving at a singular framework is extremely difficult and may not be entirely accurate. After analyzing hundreds of research papers and having consolidated our experiences working with projects, we are proposing a framework that is widely adopted and rapidly gaining industrial popularity, momentum, and support.

The industrial communication of the future

There are several enabling technologies for industrial automation – industrial protocols. As discussed earlier, protocols form the basis of standardization and are the language translators for various systems to seamlessly interact and exchange information. This section includes the core component of this chapter, connecting the physical layer of PLCs to the application layers in the cloud while focusing on advancing your Industry 4.0 implementation.

Industrial bus or Fieldbus protocols, as they are widely called, provide a digital communication link between control devices serving as a local area network. They offer important abilities such as installation flexibility, easy maintainability, and, most importantly, configurability. Real-time industrial ethernet is the base of most of the Fieldbus protocols today. Safety and production-grade readiness are pivotal requirements of an industrial environment. To satisfy both of them, timely and reliable communication is critical. Fieldbus communication provides this much-needed determinism in data communication.

Communication systems are often proprietary and varied, making interconnectivity and system integration arduous. In greenfield and brownfield implementations, the lack of standards results in extensive re-engineering efforts, often resulting in huge inertia for change. Multiple communication technologies, protocol conversion, and creating a unified data model are often the most significant challenges in any digital transformation project.

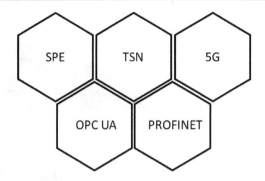

Figure 5.8 – Five tenets defining the future of industrial communication

Since the prerogative of Industry 4.0 is to have fast and reliable connectivity between and across all the subsystems from level 0 to level 5, in the following sections, we will focus on the five fundamental and advanced pillars, as shown in *Figure 5.8*. Of course, the ecosystem is broad, deep, and highly diverse. Therefore, we will endeavor to focus on pivoting toward industry standards.

PROFINET and OP CUA

PROFINET is one of the industry's leading Industrial Ethernet standards (Fieldbus). It follows the provider/consumer model for data exchange. Thus the controller and the device can only independently send data. PROFINET forms the backbone of distributed I/O and control, providing cycle times between 125 microseconds to 512 milliseconds and efficiently handling high-volume data exchange within or between machines.

The introduction of the Industrial Ethernet was the first step to addressing the challenge of digital transformation at the physical or hardware layer, but the problem of multiple protocols being used by industry still looms large. The standardization approach introduced by **Open Platform Communication Unified Architecture (OPC UA)** solves the challenges of incompatible protocols for vertical integration. It attempts to standardize **Machine to Machine (M2M)**, machine-to-MES, ERP communication, and communication to the cloud. This manufacturer-independent protocol alleviates many of the challenges of inter-protocol conversion. OPC UA provides a standard data model for over 60 industrial equipment types and supports client-server and publish-subscribe communication patterns. Robust security mechanisms are in-built with secure encryption, authentication, authorization, and X.509 key exchange.

Both PROFINET and OPC UA work hand in hand via the same physical medium to achieve two distinct results. The symbiotic relationship helps unify multiple field devices and thus improves standardization efforts.

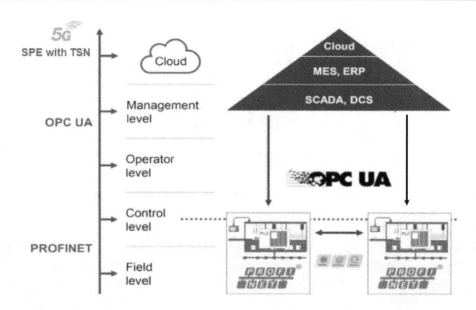

Figure 5.9 – Toward one-network infrastructure with PROFINET and OPC U/A

The deterministic real-time control data is transmitted and handled via PROFINET while OPC UA takes care of the non-deterministic device communication. This synergy is appropriately represented in *Figure 5.9*.

Single Pair Ethernet

The lower levels of ISA-95 have been the weaker links in the data pyramid of relatively aged protocols. IEEE 802.3cg, released in 2019, specified the physical layer for Ethernet communications over a single pair of conductors. **Single Pair Ethernet (SPE)** makes this connectivity automated and seamless in many brownfield environments without extensive retrofitting of the cabling or equipment.

SPE describes Ethernet transmission via a single copper cable pair. In addition to data transmission via Ethernet, SPE also supports the simultaneous power supply of the end devices via the **Power over Data Line (PoDL)**. Until now, this required two pairs for Fast Ethernet (100 MB) and four pairs for Gigabit Ethernet. SPE now completely opens up new deployment options and applications for Industrial Ethernet and Industrial IoT.

The transmission speeds of SPE can range from 10 megabits to 10 gigabits while being simple in terms of design and spacing requirements. Evolving from the automotive industry, SPE is a go-to technology in industrial automation, smart buildings, and the IoT space. Interoperability within existing frameworks is also a great advantage for SPE in adoptability.

Advanced Physical Layer

Advanced Physical Layer (APL) is an enhanced physical layer for SPE based on 10BASE-T1L. Since it adheres to IEC-60079 standards, it is considered intrinsically safe. This implies that it cannot, under any circumstance, create a spark or a fire hazard due to electrical fluctuations, which is common in industrial environments. APL has significant importance in the industrial context by providing enhanced safety from fire hazards. Ethernet APL has been developed in coherence with the OPC UA and PROFIBUS cohorts. For the purpose of this book, considering the evolution and larger shared objective, both SPE and APL Ethernet have been used interchangeably at this point in time.

Time-Sensitive Networking

Time-Sensitive Networking (TSN) has been a game changer, with its ability to prioritize and schedule network traffic, allowing time-sensitive data to be delivered to the right place at the exact right time. High-speed motion control with extreme accuracy in the case of advanced robotic control is now possible through TSN. Since accurate data transmission can be scheduled at the network layer, non-authorized data is automatically left out, resulting in increased security. TSN is here to stay in terms of extensive adoptability as it uses standardized mechanisms (independent of manufacturers), which can be integrated into the network infrastructure and end nodes.

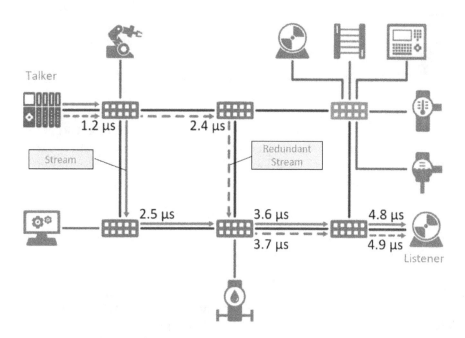

Figure 5.10 – TSN in action

The flow of information, **Talker** to **Listener**, is depicted in the preceding diagram.

Industrial 5G

The fifth generation of mobile networks, also called 5G, enables applications and services that require better reliability, improved energy efficiency, massive connection density, and lower latency. 5G is the most evolved network for the industry sector as it provides three key points, as shown in *Figure 5.11*.

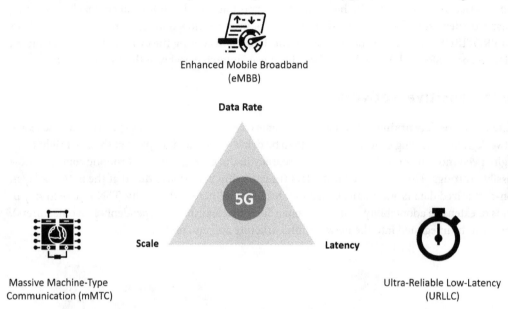

Figure 5.11 – 5G: A truly industrial telecommunication network

5G provides 10-25 times faster speeds than 4G networks through **enhanced Mobile Broadband (eMBB)**. Benefitting from the **Ultra Reliable Low Latency Communication (URLLC)** mode, 5G networks guarantee the highest reliability and lowest guaranteed latencies below 10 ms. **Massive Machine-Type Communication (mMTC)** enables magnanimous connectivity options powering up to 1 million concurrent users in an area of 1 square kilometer. 5G is designed for the industry with specialized hardware and network topologies supporting major industrial protocols. All these factors make 5G a true pillar for the next generation of factory applications.

The beginning of 5G industrial networks

When did Industrial 5G begin? Scientific collaboration in early 2020 achieved 3 ms cycle times with 1 ms latency running PROFINET and PROFIsafe over a 5G network. These extreme real-time results make use cases such as **Automated Guided Vehicles (AGVs)** and **Automated Mobile Robots (AMRs)** a definite possibility. The best aspect of PROFIsafe devices is that no unique configuration is needed while adopting 5G. For local industrial sites, the German Federal Network Agency has reserved a total bandwidth of 100 MHz from 3.7 GHz to 3.8 GHz.

This section has been an intense technical dive into the fundamental pillars of a modern industrial backbone network. The following section deals with the big picture, combining these pillars to construct a fully formed structure and strategy.

Mapping your integration strategy

In the previous section, we looked at some of the forerunners poised to shape the future of industrial communication. Each of them has great potential to alter the landscape of the ecosystem, and symbiotically, they are compelling in defining next-generation solutions, possibly leading to Industry 5.0.

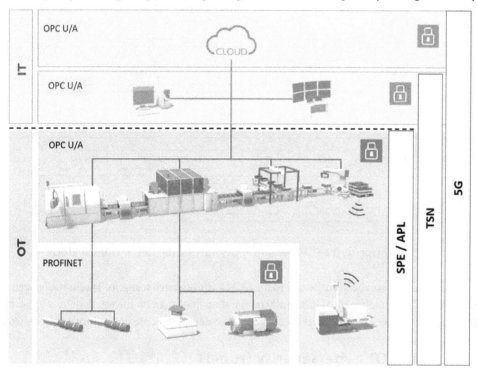

Figure 5.12 – Converging protocols

Figure 5.12 illustrates an ideal integration architecture of an industrial communication network. The sensors, machines, devices, and actuators are connected to the PLCs via a real-time Fieldbus, such as PROFINET, providing the lowest latency data transfer. The communication from the PLCs to SCADA, MES, IoT Edge, ERP, and the cloud is all unified with the OPC UA standard. PROFINET and OPC UA share the same physical layer for communication, the APL Ethernet or the SPE. They provide seamless and straightforward connectivity options on the shop floor. Time-sensitive networking makes communication robust, guaranteeing message delivery and addressee correctness. All of this can then be powered through Industrial 5G networks. Thus, we have a converged ecosystem in place.

Figure 5.13 represents the various layers of the pillars mapped onto the **Open Systems Interconnections** (**OSI**) model. Multiple components of SPE, APL, TSN, PROFINET, OPC UA, and 5G are superimposed on the same OSI structure to visualize the juxtaposition.

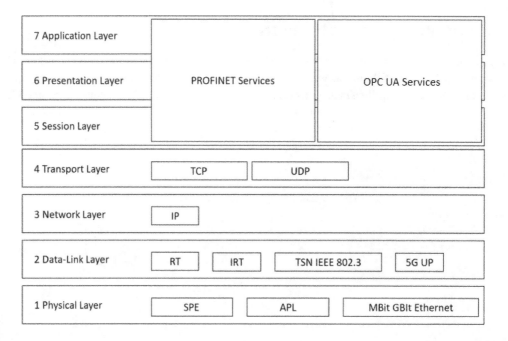

Figure 5.13 – Mapping the OSI layer with PROFINET, OPC UA, TSN, SPE/APL, and 5G

Another area for review is cybersecurity due to the OT/IT convergence scenario. Traditionally, security has been perceived and implemented differently on the shop floor than on the top floor. It is important to understand and appreciate the differences if you are to design a secure and future-ready network.

Differentiating OT cybersecurity from IT

There are quite a few differences between cybersecurity implemented at the OT and IT layers, even though the underlying technologies are similar. An OT system needs to maintain defined availability to reach critical production targets. This implies that a factory operator is unwilling to see a flash on their screen saying *automatic updates are in progress; the system will restart in 5 minutes* while performing production operations. Production comes first and cannot be jeopardized. Security controls need to work in tandem with the production schedule.

OT systems control physical devices and are often influenced by legacy and modern technology. The implementation of a one-size-fits-all approach will never work. The shelf life of IT systems is relatively short in comparison to OT systems. Due to the expensive nature of OT systems, they are also not easily replaceable or upgradeable. Thus, the same patching strategy and frequency would not work.

Starting with user access to the setting up of DMZs and multi-firewalls, there are multiple vital topics surrounding OT cybersecurity implementation. OT security best practices are listed in *Figure 5.14*. We shall examine this more in *Chapter 6, Enabling Industrial IoT,* and beyond.

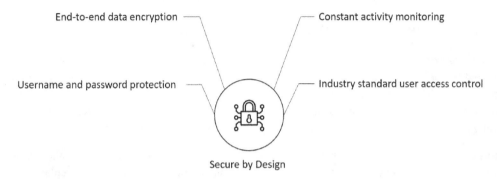

Figure 5.14 – OT security best practices

Major OT cybersecurity threats

In 2021, the meat processing giant JBS was attacked through ransomware, disrupting the distribution of meat products to millions worldwide. The organization had to shell out 11 million dollars to restore normal operations.

A major automotive company's global operations were halted in June 2020 due to a Snake ransomware attack at one of the plants. The ICS system was reportedly hacked, and important data was encrypted.

A pharmaceutical company was subjected to a ransomware attack in 2017. The malware got into over 30,000 computers and 7,500 servers. Years of research were lost, and normal operations were seriously impacted.

Triton was one of the first OT-focused attacks that resulted in operations mayhem in a petrochemical plant in 2017. It targeted vulnerable systems with predictable or weak security, targeting **Safety Instrumented System** (**SIS**) controllers, a critical component of industrial processes. Compromising this security layer could have resulted in physical hazards as a consequence.

In 2015, a Ukrainian power producer was attacked by a team of hackers, likely a nation-state. The attack led to over 225,000 customers losing power for 6 hours.

Stuxnet is one of the most famous cyberattacks ever in the industrial sector due to its sophistication and impact. It was one of the first malware files designed to attack an industrial system, which happened to be a nuclear plant in Iran. The virus got into the facility and changed the running code in the SCADA system that controlled the centrifuges used to enrich uranium, with the intent of making them operate in a way that would lead to failures.

Having obtained a keen understanding of the integration strategies and associated nuances, the next logical step is to look at the convergence of IT and OT systems. This convergence is vital to creating a unified framework that will be the conduit to running futuristic applications.

Where IT meets OT in industrial systems

The cornerstone of the Industry 4.0 digital transformation has been around cost reduction, worker safety, rapid technology adoption, end-product customization, fast product delivery, just-in-time, high quality, and minimal waste in production. The IT/OT convergence enables the same transformation paradigm. Data that was originally transactional from the factory floor to the executive suite is now becoming real-time and empowering corrective decision-making at the same rate.

Peer-to-peer, or M2M communication, with collaborative functioning and optimization goals, is now becoming a reality. This is because machines, the single valuable component, are getting replaced by data. This, in turn, is now helping to operate machines more efficiently.

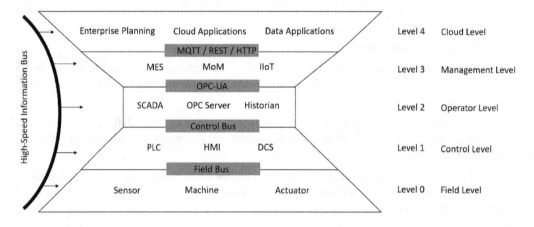

Figure 5.15 – Evolving ISA-95 architecture from a triangle to a connected web

The hierarchy of the data pyramid is now dissolving, with the top layers expanding in scope and data interchanges happening from and to multiple layers almost seamlessly, as depicted in *Figure 5.15*.

Why do we need convergence?

As a result of our understanding throughout this chapter, it is now inevitable that IT and OT systems need to converge into a single interconnected structure. Real-time data transfer from various subsystems at the manufacturing level is integrated into what we can consider a unified data fabric. Bolstered by data context, historical metrics, and machine learning models, analysts and management can create an entirely accurate picture of the current state and identify corrective measures to avoid a potential failure state.

Armed with this knowledge, corrective actions are again sent back in real time to the manufacturing floor, enabling and empowering people for faster and better-informed decision-making. This can help to alter system behavior toward higher performance, efficiency, and minimized errors. Hence, a converged ecosystem is primordial to enable this workflow.

What drives IT/OT convergence?

IT and OT have traditionally shared varied and disparate goals, as shown in *Figure 5.16*. While the plant always desired low latency and safety, the IT systems are designed around cost optimization and protection from cyber threats. Efficiency and repeatable performance are a high priority on the shop floor, while scalability and agility are a priority on the top floor. **Industrial IoT** (**IIoT**), as we shall see in depth in the next chapter, is driving this convergence. IIoT brings the power of connectivity, interoperability, and intelligence, which helps bridge IT and OT systems.

The individual goals on either side become shared goals of the entire system. This becomes a combined solution as we evolve toward an integrated manufacturing ecosystem. The playing ground gets muddled with brownfield implementations and thus drives our case for standardization. Creating interoperability through retrofit mechanisms and building data conduits through IIoT is critical in such scenarios. We will analyze this further in *Chapter 6, Enabling Industrial IoT*.

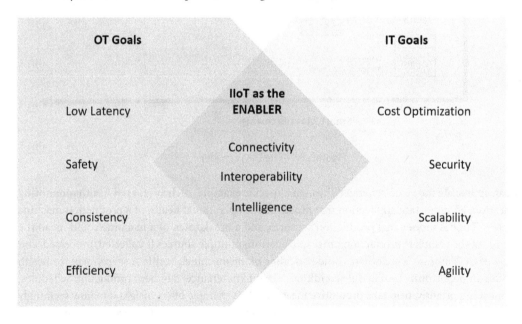

Figure 5.16 – OT and IT goals with IIoT as the enabler

Let us now revisit the DIKW model and summarize our understanding of it.

Revisiting the DIKW model

In *Chapter 3*, *In-Situ Environmental Monitoring*, we discussed, in brief, the DIKW model or the *Data, Information, Knowledge, Wisdom* model. We talked about the importance of the collection of data and how it transforms over time into wisdom, combining context, history, and patterns, given time and reasoning. The exchange of data without context has little value. *Figure 5.17* depicts the various stages of direct value addition in a connected ecosystem.

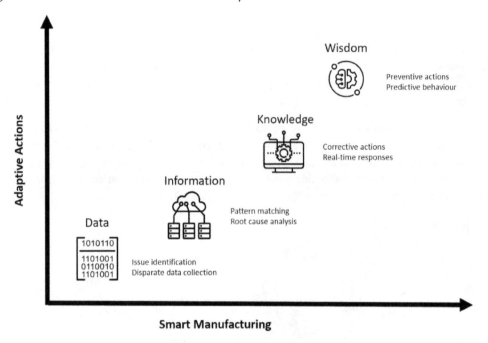

Figure 5.17 – The DIKW model

We can appreciate the value generated through a specific example, such as an asset health monitoring application, an important application that monitors the operational health of a production machine remotely. Time is money in a production scenario, and a breakdown of a machine could mean the stoppage of the complete production process. Data from multiple sources is collected, processed, and correlated to determine a quantitative understanding of the machine's health. A snapshot of the health provides an opportunity to schedule condition-based maintenance, augment production schedules, and, most importantly, help take preventive measures. This example offers insight into how seemingly unrelated and binary values can be transformed with context, history, and intelligence to create value. Consider the progression in *Figure 5.17* as you traverse through this data transformation journey.

Figure 5.18 provides a visual understanding of our asset health monitoring use case. The PLC controlling the machine directly streams the binary or ON/OFF data from a piece of equipment. When the machine sends non-operational data during its scheduled operations, it means that there is an anomaly, and the availability of the machine is reduced. When we obtain additional data from various sources, we can understand that there appear to be some anomalies.

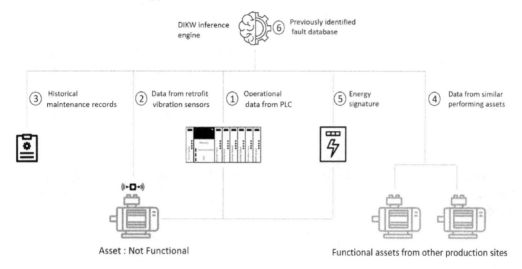

Figure 5.18 – DIKW model through an asset health monitoring use case

Let us combine the PLC data with the data from a retrofit vibration sensor, energy meter, and parallel lines. We now have data from similar performing lines, the energy signature of this machine, which directly relates to the load on the machine before the anomaly. Add in the historical maintenance schedule of the nonperforming asset. When data from multiple sources is correlated, pattern matching helps ascertain the root cause very quickly. We can then map it to a standard anomaly list and identify the reason and the corrective fix. Thus, the knowledge base is built with methods to fix the asset. When applied to a machine learning model and fed with real-time performance data, this knowledge base can predict future anomalies and prescribe corrective measures.

Adding context through metadata and making the data available in real time for dynamic decision-making is crucial. Maturity in smart manufacturing through adaptive decision-making lies in the free flow of information horizontally and vertically. The IT/OT convergence enables us to achieve this possibility and the IIoT layer, with the data fabric as the driving force behind this transformation.

Summary

This chapter started on a historical note, looking at the evolution of industrial automation systems. We talked about PLCs, DCS, and SCADA systems. As we undertook a journey to understand our various manufacturing ecosystems, it helped us appreciate the need for sophisticated and purpose-built control systems to optimize production. The evolution of the PLCs has been fascinating, and the industry keenly anticipates the next significant set of advancements and milestones.

We then investigated a brief composition of a PLC, its types, architecture, and programming constructs, before comparing them to microcontrollers. Industrial protocols are essential for a successful Industry 4.0 implementation, especially in a complex heterogenous multi-player and multi-generation ecosystem. We discussed the details of some of them in this chapter.

The combination of PROFINET for real-time acyclic communication and OPC UA as a standard for vertical cyclic communication over SPE on the fundamentals of TSN, bolstered by Industrial 5G, is a single-line takeaway for this chapter.

Creating a solid integration strategy is one of the key steps in implementing a digital transformation program at a manufacturing facility. To achieve this, we discussed various approaches and best practices to help build the connectivity data fabric.

The next chapter will focus on IIoT, the fundamental glue enabling IT and OT convergence. This will provide us with practical examples of IT/OT integration, pitfalls, and strategies to overcome hurdles during real-world implementations.

Enabling Industrial IoT

The integration of existing brownfield environments with more modern technology is a significant challenge for industries all over the world. **Operational Technology** (**OT**) systems such as SCADA that manage and control operational efficiency and safety are used to control real-world devices and systems. The challenge is to integrate these systems with IT systems and networks, allowing data and control to flow beyond the control room. Our primary responsibility is to ensure that production systems are not impacted, and integration points must be carefully specified and standardized.

This chapter marks the book's halfway point and logically forms the crucial mid-section or the integration layer for enabling industrial **Internet of Things** (**IoT**) applications. The objective is to dive deep into the complexities and nuances to appreciate the need for convergence between IT and OT systems. We would also look at interesting case studies and implementations of the digital twin, which is a direct consequence of and possibility within IT/OT data convergence. While readers will be aware of the challenges of integration in a typical manufacturing scenario, they will also take away strategies and ideas to enable such convergence at the end of this chapter.

In this chapter, we are going to cover the following main topics:

- The opportunity of Industry 4.0

- Understanding IT and OT convergence

- Architecture decisions for enablement

- Adopting technical innovations

Technical requirements

This chapter will benefit from a good grasp of IT and OT technologies and services, being a core chapter for architectural studies. An overview of solution architecture will be an added advantage. Although we will only go over each of them quickly and how they interact, we will also be leveraging services from AWS, such as IoT TwinMaker, IoT Analytics, and so on. Therefore, a good understanding of the AWS ecosystem will help readers appreciate the chapter flow.

The opportunity of Industry 4.0

One of the major concerns of industries worldwide is the integration and continued growth and innovation of existing factory environments with modern industrial practices. The control of real-world devices and systems (OT) are often closed-loop systems designed to manage and control operational efficiency and safety. Integration of these environments with enterprise IT systems and networks can be challenging. Integration points must be identified, defined, and standardized to ensure effective connectivity and minimal impact on these often long-running production systems.

Sensors, motors, gauges, and other intelligent electronic devices, such as **Remote Terminal Units (RTUs)** and **Programmable Logic Controllers (PLCs)**, as well as **Human-Machine Interface (HMI)** control system assets, are examples of OT assets. Modern PLCs and controllers have IT integration built in using a variety of protocols; however, this is not always the case, or the devices may not be as open as we would like. We will also need to discuss options and protocols for legacy systems that may use proprietary or less well-known data control protocols, especially serial protocols that are very tightly designed.

Despite years of interaction between OT and IT networks in an industrial setting, viewpoints and techniques remain disparate. It begins with organizational responsibilities in the manufacturing environment, such as production versus IT network ownership, which is a problem that is particularly prevalent in large businesses. Furthermore, each side brings different experiences with industrial technology, such as PLCs, I/O systems, sensors, or software systems, to the table.

From a strictly IT perspective, network administration and control have a lot of baggage. Cybersecurity, data integrity, usability, and serviceability are the top priorities in IT. To give some perspective, the smooth operation of OT equipment and systems is a top priority for plant and operations management. Errors, particularly failures, can have immediate and significant repercussions for production system availability and functional safety. Layer too much network control on top of this, or more distinctly, network controls that are not well managed, and finger-pointing can become prevalent.

OT and IT employ different terminology, from network tiers to communication protocols. In the OT industry, for example, complete manufacturing lines, including network solutions, are frequently built and delivered using components from **Original Equipment Manufacturers (OEMs)**. Digitalization projects such as Industry 4.0 and the Industrial IoT amplify these discrepancies. Instead of coexisting in separate, parallel universes, it is in both OT and IT's best interests to speak the same language and communicate more than ever before.

Collaborating on integrating equipment, operation, service, and maintenance requirements for industrial networks is diverse and often specialized. As a result, things usually cannot just be plugged into converged higher-level IT networks. Physical as well as logical isolation of OT and IT networks may be required by relevant security standards and guidelines, as well as to maintain operational effectiveness as systems evolve. That is why coordinated teamwork and a high-performance, secure network transition are more crucial than ever, but in a way that allows OT to feel comfortable about the effort.

IT/OT interconnections aim to move data from the OT world into the light, allowing better visibility, control, and management of plant operations. Every change attempt should begin with a characterization of the desired outcomes (right to left). *Figure 6.1* visualizes the flow from how desired results eventually tie back to the data being collected.

Data   Insights   Actions   Desired Outcomes

Figure 6.1 – IT/OT interconnections

It is important to try and follow industry best practices when preparing for integration. As mentioned in previous chapters, The ISA-95 standard is widely applied for IT/OT integration, which views IT/OT landscapes as five-layer stacks, which is how providers often explain their solutions. One caution: don't let standards or reference architectures become anchors wrapped around your neck. References are a great starting point, but that is all they are, designed to be adapted and shaped to your environment and needs.

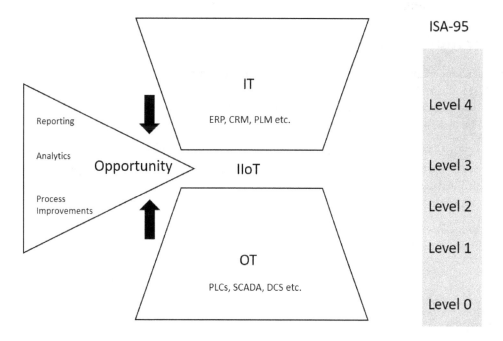

Figure 6.2 – The opportunity for optimization lies with IT/OT integration

Figure 6.2 depicts a common beginning point for most IT/OT integrations for many of these initiatives and an efficiency gap, which is actually an opportunity. Organizations constantly include IT and OT data in their broad reports, which are rarely tailored for gaining insights valuable to day-to-day operations. Convergence value is driven by identifying and then addressing the efficiency gap. This has always been a helpful approach to thinking about convergence initiatives.

There is also an opportunity to consider the kind of infrastructure we should consider while carrying out integration – should you take everything to the cloud data lake, for all analytics? Or do you preprocess data at the edge, only sending key results or flags and outliers to the cloud? Cloud platforms make data analytics more cost-effective, adaptable, and secure—when working with IT data. However, as OT data is added into the mix, these advantages may disappear and be replaced by disadvantages. Cloud analytics systems can be hampered by large amounts of real-time, time-series data—precisely the type of data generated by OT that could push up the cost of hardware subscriptions. In *Chapter 10*, *Intelligent Systems at the Edge*, we will also learn about hybrid solutions and high-speed data acquisition at the edge.

While we firmly believe that the opportunity for optimization and a robust architecture design lies in the holy grail of IT/OT integration, the process is complex due to the simple fact that these two systems have evolved from different ecosystems, purposes, and principles. Let's explore them in detail in the following section.

Integration complexity

The backbone of any digitalization effort is providing a secure data interchange between OT and IT. Most industrial networks use BUS, Ethernet-based, or wireless connectivity, which can differ in several ways. Each network has its own set of needs, to achieve a reliable data interchange in enterprise networks, these needs must be thoroughly analyzed and considered when planning the OT/IT link.

IT and OT can no longer be separate for organizations to compete effectively. However, bringing them closer together will result in increased complexity. Equipment operators and plant engineers are confronted with a growing wave of issues as expenses have risen, cyber and operational risks have increased, workloads have climbed, and compliance has become more complex.

A deeper breakdown of some of the differences between IT and OT can help us understand why they are sometimes looked at differently. *Table 6.1* provides a simple analysis of the differences between the two.

	OT	IT
Environment	The OT network is mostly industry-oriented and communicates with equipment.	Business-oriented IT networks deal primarily with information rather than hardware.
Data	Monitoring, control, and supervisory data are examples of several types found in OT networks.	Transactional, streaming, and bulky data are all sorts of data in IT networks.
	The OT network is based on real-time data processing.	The IT network processes data in a transactional manner.
Availability	End-of-life situations might arise because of OT network failure.	Data loss can occur if an IT network fails.
	Any disruption in the OT network will have a direct impact on the whole business.	IT network failure can negatively influence a company's credibility, irrespective of the sector.
Access	The OT network controls physical access to any device.	The IT network controls the authentication of devices and users to the network for security.
Patching, Maintenance	Only during operational maintenance windows does the OT network require network upgrades.	Network upgrades are expected in IT networks.
	The vendor is very often responsible for patching through remote access or by minimum audibility.	The IT team or department owns changes.

Table 6.1 – IT/OT differences

One of the areas of concern is a lack of visibility. After all, what you can't see, you can't support and defend. Existing security solutions are designed to address this on the IT side, but on the OT side, it's a different story.

There are many differences between IT and OT operations simply due to their evolutionary nature and prioritized value performance indices. The corporate IT team, led by a CIO, is responsible for information technology –computers, applications, networks, and telecommunications equipment. Plants and manufacturing focus on OT, driven by manufacturing and production, owned, operated by the factory engineering team, and designed to run machines and facilitate industrial automation within the local environment.

An example of a possible conflict would be the case of security patching and system OS updates. IT would roll out updates as soon and frequently as possible. The roll-out process is relatively straightforward and systematic as IT assets are well-regulated and audited. On the OT side, the ecosystem is entirely different. Non-standardized computers run a variety of operating systems to support legacy applications.

Updating any of those would often mean the requalification of machines and may take several weeks or even months depending on the complexity and availability of application support.

Time-series data

Time-series data is another feature of OT that needs to be considered. It's an integral part of any IT/OT integration project and is a distinct type of data from what is frequently found in commercial IT systems. It's also time-variable, which can add some complexity to managing and using the data. We must interpret data as is by adding contextual information. Professionals would have no idea whether a temperature graph that shows *59°F* is good or not if they don't know anything about the equipment that created the data point (is it a cooling unit or an oven?). The local operator of the equipment knows what the ranges should be, but for higher-level analysis, we need to include more information—precisely, metadata such as what specifications or conditions the machine has for normal operations. We usually need good data quality management to increase contextual metadata while also supporting data governance.

In the age of the IoT, IT/OT integration is critical for deeper understanding and the overall management of these systems and processes. Because of the inherent interoperability limits of OT, as well as remarkable developments in (usually lower cost) IT, general-purpose IT is becoming much more widely accepted within industrial operations. This is true even for some of the most critical industrial applications – in fact, the criticality of these environments demands better instrumentation and visibility.

Let us now venture into the *why* part of the convergence and its explicit and implicit advantages.

Understanding IT and OT convergence

IT/OT companies are developing to keep up with the growing usage of data analytics. Executives are more aware that data may help them make better decisions. IoT advancements have enabled physical machinery and the virtual world to merge, generating massive amounts of data that can be used to visualize OT operations and identify areas for improvement. Companies can better integrate OT with business systems and **Enterprise Resource Planning** (**ERP**) with detailed data on industrial functions. OT is increasingly being called upon to interact with complex systems and networks that have traditionally been the domain of IT. **Product Lifecycle Management** (**PLM**) in the manufacturing industry, for example, is being disrupted by the usage of *digital twins*.

IT/OT convergence, aided by IoT and big data, can result in cost savings and improved performance, productivity, and agility, as shown in *Figure 6.3*. Convergence of IT and OT also means greater visibility across the supply chain and more information on the status of OT, which can help managers uncover and fix inefficiencies. Manufacturers and service providers must adjust to continually altering consumer needs as digital gadgets reshape consumer expectations. Cost savings will be the most significant benefit of IT/OT convergence in many circumstances. IT can assist OT teams in aligning with larger business models and driving management system architecture. On the other hand, the OT knowledge of physical processes and machinery can help perfect the IT design and vision within the manufacturing process.

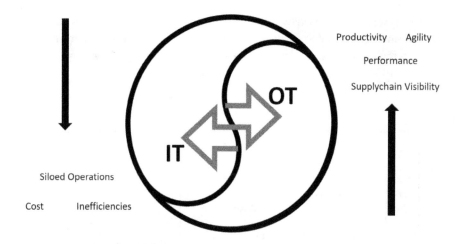

Figure 6.3 – The convergence advantages

Enterprise systems can advise OT operations to minimize production expenses and adjust inventories to changing consumer needs rather than cut costs by arbitrarily cutting staff or materials. Accepting that putting IT and OT together has its own set of obstacles is a necessary step to reap these benefits. True IT/OT convergence necessitates the digitization of OT and the development of systems that turn data into action. Departments that have historically functioned in silos, on the other hand, may find it challenging to interact efficiently. Despite IoT breakthroughs and provable benefits, IT/OT convergence continues to come up against considerable organizational opposition.

IT and OT divisions in many firms work in diverse ways, with different approaches and aims – not only that, but they also frequently work on incompatible projects. IT and OT staff may even disagree on the company's ultimate direction. However, to reap the full benefits of IT/OT partnerships, IT and OT leaders must actively support the convergence effort. IT/OT convergence necessitates more than a vague mandate to collaborate; it necessitates a clear vision and open lines of communication.

The following few sections are going to be exciting and are designed in a way to appreciate the importance of IT/OT convergence. We will explore the concept of digital twins derived from an asset administration shell. The asset administration shell is a complex topic and will be discussed in *Chapter 9, Taking It Up a Notch – Scalable, Robust, and Secure Architectures*. Digital twins, in simple terminology, can be defined as virtual representations of the physical entity in real time or near real time. Without convergence, pushing data from the field to the cloud, this concept would not exist. With this context, let's explore the facets of the digital twin.

The emergence of the digital twin

Digital twins have emerged relatively recently, with mixed reactions in the industry. As usual, the marketing machines of many IoT solutions have twisted them into the final solution for all your problems.

Thousands of IoT sensors are now used by manufacturers to track machine operations using various metrics. This data can be combined to create something called a *digital twin*. A digital twin is a high-fidelity data-driven replica of a machine's physical assets that constantly interacts with fresh data.

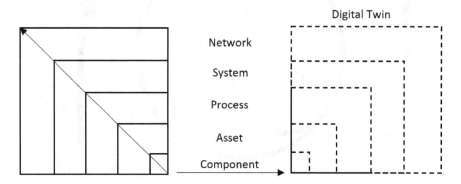

Figure 6.4 – Types of digital twins

Digital twins are thus a real-time representation of the behavioral performance of a component. Digital twins, on a basic level, can consist of a replica of a simple element, such as a car suspension, modeling its on-road performance. Multiple elements holistically form an asset digital twin, abstracting and aggregating various data points, for example, a transmission system. Process digital twins can also be a sequence of interdependent activities of one or more assets in conglomeration – for instance, a production line. The system is a higher abstraction of a unit or entity, while network digital twins depict complex interactions between multiple systems in a similar or different environment. This is illustrated in *Figure 6.4*.

These digital twins assist manufacturers in actively monitoring and evaluating the field performance of equipment. For example, digital twins can be used for predictive maintenance to determine when gear needs to be serviced or replaced. Many OT service providers are quickly forming alliances with IT-centric providers to take advantage of digital twins and other IoT capabilities. Specifically, they can provide some of the following benefits:

- Store metadata about your devices and equipment

- Provide the latest conditions or state of your equipment

- Combine both on and offline information, metadata, and external data to provide a holistic picture of the device, status, or maintenance plan

- Apply physics-based models and statistical data models to simulate, test, and validate the performance peak

You can see how this can be pretty powerful. With out-of-the-box IoT, we can see things such as **Revolutions per Minute** (RPM), energy use, flow rate, etc. This is useful and powerful information.

Now, combine this with the make, model, size, capabilities, and maintenance information, and this gets you as close to the machine as possible without standing on the shop floor. From the perspective of an AWS user, a powerful service called IoT TwinMaker exists. Let's now look at an introduction to this service and its capabilities of interest, with various applications for your chosen domain.

AWS IoT TwinMaker

Building a functional digital twin is quite hard. It requires a lot of work to model the physical equipment and connect the two to allow the twin to represent the state of the actual machine. It is even more work than using that model or twin to perform helpful analysis or what-if scenarios before something is done to the device itself.

AWS provides the concept of a Device Shadow, which allows you to define JSON documents outlining a device's current state and metadata. Using **Message Queue Telemetry Transport** (**MQTT**) topics, the user can get, update, or delete information about a device. It is a useful abstraction of your device but slightly limited in how you interact with the equipment. Allowing you to combine states and metadata can provide a more complete picture of the equipment to external systems or users in real time.

More recently, AWS has provided a much more comprehensive solution called **AWS IoT TwinMaker**. TwinMaker allows you to build detailed models of your entire plant or factory, including all the equipment and components inside. Combining 3D models of the factory or equipment with current data provides a full view of the environment, measurements, and production output.

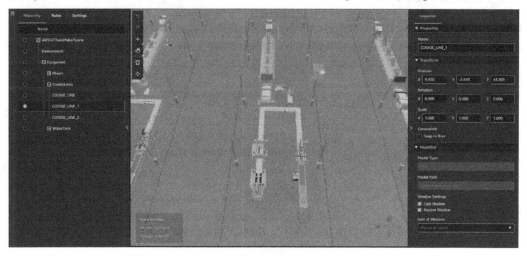

Figure 6.5 – AWS IoT TwinMaker scene composer

Entities can be built out to represent elements in your physical environment. These entities can be connected to data systems: either internal data systems within AWS or external data systems. *Figure 6.5* illustrates how the TwinMaker scene composer view can be used to combine models, sensor data, and video from your environment to provide a holistic view of your factory floor.

The next section circles back to the book's core, enabling you, the reader, to take informed architectural decisions toward a best-fit solution. We will explore data aggregation options and mechanisms for ingesting data into the cloud. All of this will, of course, be focused on services provided by AWS.

Architecture decisions for enablement

This book is about making good architectural decisions while designing and building your industrial IoT systems. There is a lot to understand about why these decisions are important, so we spend a lot of time talking about the benefits of Industrial IoT and Industry 4.0 since the technology would be a waste of time and money for its own sake.

Often, we don't know our end game; we know that technology and its potential benefits are passing us by, and we need to catch up to stay in the game and stay relevant within our industry. This is OK and trying something is better than doing nothing, but better still is thinking about your industry, your constraints, and the environment, and then designing something that will allow you to achieve your end goals most effectively. We will now explore AWS's different IoT-related services and focus on some of these in subsequent sections for the exploratory data ingestion problem. This will enable the readers to make informed choices on the thought processes behind applying specific methods while architecting solutions.

AWS services for IoT

At the time of writing this book, there were about 13 IoT services within the AWS catalog broken down by area. We have outlined the currently available services in the following table:

Device Software	
FreeRTOS	An open source real-time operating system for microcontrollers
AWS IoT ExpressLink	Hardware modules that can be added to consumer devices for simplified IoT-based product development
AWS IoT Greengrass	Open source runtime for building and deploying intelligent edge applications
Control Services	
AWS IoT Core	Securely transmit messages to and from IoT devices and applications
AWS IoT Device Management	Register, organize, monitor, and remotely manage connected devices at scale
AWS IoT Device Defender	Managed service to secure your devices and audit device configuration and behavior
AWS IoT FleetWise	Collect, transform, and manage vehicle data in the cloud

AWS IoT RoboRunner	Infrastructure for integrating robots and building robotics fleet management applications
AWS IoT 1-Click	An AWS service that enables simple devices to trigger AWS Lambda functions that can execute an action using supported devices – imagine a button to summon assistance or a ride
Analytics Services	
AWS IoT Analytics	Managed service that automates the steps to collect, filter, transform, and analyze IoT data
AWS IoT Events	Managed service that allows you to detect and respond to events from sensors and applications
AWS IoT SiteWise	Managed service that allows you to collect, organize, and analyze industrial data at scale
AWS IoT TwinMaker	Build and use digital twins for real-world systems

Table 6.2 – AWS IoT services

This extensive list continues to grow to cover new use cases across the IoT spectrum. This list does not include many additional services that are not specifically IoT. Other options may be used to bring data into the platform, such as Amazon Kinesis. Still, it does give us a view into the complexity and variability of the choices we need to make as our architecture evolves.

In the next section, we will explore three AWS-based data acquisition options: one based on IoT analytics, one using a custom lambda function, and the other employing Kinesis Data Streams and Firehose. The idea is to present multiple avenues and methods to the readers, but choosing an option or a potential combination definitely depends on the use case at hand and the system design parameters.

Revisiting cloud data ingestion

We covered a lot about data ingestion in *Part 1* of this book, but the topic is large and varied as you are learning, and we want to look at this from the perspective of making strong architectural decisions.

Figure 6.6 illustrates some of the options discussed in previous chapters around data integration and processing from an IoT environment. This is a simplified view of the options designed to help you understand some of the choices you might make depending on your environment and goals.

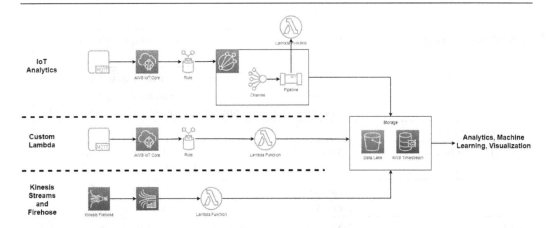

Figure 6.6 – IoT data processing route options

The preceding figure defines three main paths that can be undertaken. In reality, you can choose a combination of these services and process your data in various ways through the intake pipeline. There are additional options we can pick, perhaps by ingesting data through a cluster of Kafka instances or using another AWS service or approach. Still, the goal is not to overwhelm you with too many options and force you to make one correct choice – instead, to help guide you in making a practical decision that maybe needs to be adjusted as your environment evolves. Let's look at these three options and discuss each in turn and how we might choose each.

IoT Analytics

This is an excellent first choice. It allows you to start quickly, but some work still needs to be done. The service guides you in getting data into a structured format in a way that follows current best practices, so if your data or environment is set up in a way intended to take advantage of this approach, you can benefit greatly.

This is an AWS-managed service, so there is no need to worry about servers or instances and less about scaling as you build new channels and pipelines to define, split, or process data. There is still the need for lambda functions in probably every case, so enhancing or decoding data requires some manual integration to ensure data is interpreted correctly.

In some ways, IoT Analytics is not great for data that comes in wirelessly, such as using the LoRaWAN protocol. This data is encoded, and the JSON contains far more information than is necessary for IoT analysis, so it doesn't fit precisely into the Analytics mode of initially channeling data into a raw data lake storage as cleanly as other options. Your raw data ends up encrypted, which we don't love. This might end up being a religious choice for many architects.

Custom Lambda

There is nothing particularly special or exciting about this approach, but it is probably one of the most used approaches by AWS customers to date. It provides the most control and ability to manage your intake pipeline, but also the most effort to build and continually manage your data stream as it evolves. In most cases, these lambda functions are necessary in any case to modify or enhance your data packets as they are processed. Additional features provided by managed services can easily be added.

We provided a deep dive into this approach in *Chapter 4, Real-World Environmental Monitoring*, where we wrote a lambda that mimicked the features of an IoT Analytics pipeline. There is no question here of functionality. This approach allows you the most flexibility if you know what you are doing and are willing to take the extra effort to code this functionality yourself – but another question is scale. Is this the right choice for big data integration, where data may flow several times a minute or at sub-second intervals?

IoT versus big data

Sometimes, these words are used together, so it is worth discussing the difference.

Big data stems from the explosion of data gathered by organizations over the last several decades. This data and its collection rate continue to grow, and big data is a concept or approach designed to allow corporations to manage, secure, and learn from the information that these data stores can provide. **IoT data** is more specifically tied to physical measurements of systems or the environment. IoT refers to *things* that can be machines, vehicles, and even people.

Big data may or may not contain IoT data. Most likely, some of this comes from IoT streams and can be used independently or combined with other streams to enhance its value. Big data techniques can come into play if IoT data has enough volume or velocity.

For this large-scale data integration, you may need to look at additional services beyond the AWS IoT services.

Amazon Kinesis

Amazon Kinesis is another fully managed service that provides streamed data processing capabilities. The service also includes the power for video stream analytics, which we will not go into in this chapter. One of the core capabilities of Kinesis is the ability to process data in real-time and scale that processing as your data needs grow.

Kinesis Data Firehose is an additional feature that allows you to transform, package, and store your data in an AWS data store, such as an S3 data lake. Firehose can package your data based on your needs on a scheduled basis, such as every minute or hour. These data packets can be bundled, transformed, and stored to allow them to be used further upstream for analysis. With all this capability comes increased costs, so if that is a concern, you should understand what those costs will be and how to manage them effectively.

We have discussed a few options to help you understand the approaches you can take to process your incoming data. You may want to consider the volume and size of data and the cost of processing. Using Amazon Kinesis can be more costly, but if you need it to process your data correctly, use it. If you want to use it because it's cool, then you may want to evaluate your costs against the benefits more closely to see whether it's worth the long-term cost. Understanding that each option provides a different cost-to-benefit balance based on your data processing needs is essential in providing good value to the business. Next up are the available options for data collection or acquisition.

Data collection options

When getting data from devices and equipment, there are generally a couple of options to consider. Often, depending on the data you need to collect, it may be a combination of options, as shown in the following figure.

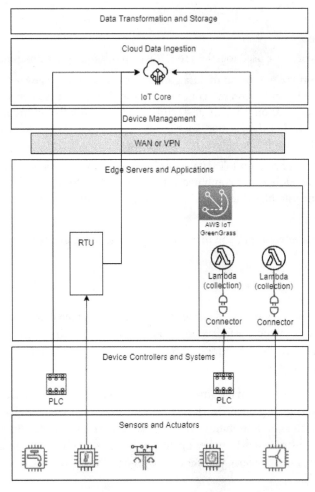

Figure 6.7 – Data collection options

We can categorize these options in different ways, and, of course, this is an oversimplification of categories, but it's a starting point when thinking about options and courses of action. We have three different options for data acquisition: one is from the PLCs with connectivity ports, the second is from an edge computer to which all the field devices are connected, and the third is using in-situ intelligent sensors, which were discussed in detail in *Chapter 3, In-Situ Environmental Monitoring.*

Process or control devices

Many PLCs and control systems have built-in data enablement. Most modern PLCs have multiple ways that data can be received and transmitted to the controller in meaningful ways. There are two significant variations of PLC:

- The fixed or integrated PLC has all input and output options built into the controller and is defined by the manufacturer.

- The modular PLC is generally rack-mounted and contains a CPU and several I/O modules that can be connected to provide connectivity. This type typically has more options for collecting and viewing data outside the PLC network.

Many PLCs, especially from the older generation, communicate only with specific protocols, such as Modbus or Profinet. This communication may be provided over a serial interface such as RS-485 or an Ethernet connection as the protocol requires. One option for integration is to provide an adaptor to interact with the PLC in its native language and transfer the data to something more useful to our IoT activities. One example is using a Modbus-RTU protocol adaptor to pull data from the PLC network and send the data using MQTT.

Some PLCs have built-in options, such as web servers or even built-in IoT capabilities. These options are a no-brainer for new factories and environments and can provide out-of-the-box IoT capability for your systems. However, the cost and effort to replace all your systems may be prohibitive for brownfield environments. In the next chapters, some of these options will be explored and demonstrated in more detail.

Cloud-enabled edge devices

Edge computing has more recently become the new savior of the manufacturing world. For many reasons, edge computing offers a lot of manufacturing capability beyond just collecting and forwarding sensor and system data. With cloud-enabled devices, we are focused on computers placed within the factory environment (on the edge) and can work with data before it is sent over the internet to the cloud. This opens a lot of doors for us, such as the following:

- We can use an edge device in its simplest form to connect with PLCs and forward data to AWS

- We can locally review data on the edge and make decisions about what to do in certain conditions

- We can aggregate, average, or summarize data to reduce the amount of data leaving the factory setting

- We can perform more advanced calculations or machine learning on data to identify problems or issues closer to real time

These are just some of the options open to us with introducing edge computing into our environment. AWS **Greengrass** (**GG**) provides an edge computing runtime environment to provide all of these capabilities. We will dive much deeper into GG here and in future chapters. Know that there are many options for edge computing available today.

The whole industry has evolved on helping to capture, analyze, and transmit your industrial data, making capturing data as simple as possible. Edge data processing is only as good as the data it can collect, so there is still the need to communicate with PLCs or other equipment using adaptors or possibly more recent protocols. This is our big adventure – understanding what is available in our environments today and what is the best way to approach each silo of data.

In-situ sensor placement

We reviewed this in detail in *Chapters 3 and 4*: however, we focused on wireless sensors in out-of-the-way locations, placing sensors in isolation. These sensors connected directly to the cloud via wireless protocols and sent data to be processed based explicitly on one or maybe two measurements.

In a more traditional industrial sense, we may not have the correct sensors on machines that we would like to have – to measure temperature, for example, or voltage on a machine. Many legacy pieces of equipment may not have those types of measurements available. A typical example may be measuring pressure on a press and then later correlating that with product quality.

There are probably many cases where operations would like to understand some aspect of the equipment they are running. Those conversations would be an interesting place to start your journey, allowing you to focus on providing almost immediate value to key organizational stakeholders.

Having looked at the options for data acquisition, let us spend some time exploring the capabilities of AWS IoT GG, which powers many intelligent edge use cases across industries.

Intelligence at the edge with AWS IoT GG

Edge computing is an interesting proposition. The marketing hype around it is tremendous, but it's not a simple decision. It would be best if you started with the use cases around what you want to achieve and why that cannot be achieved with traditional or cloud computing. Determining the *what* and *why* also provides better guidance on the *how*. Processing data or machine learning at the edge is helpful for some fundamental reasons:

- You need immediate feedback and control. If you suspect some issue or your machine learning models detect an anomaly, you want to address that as close to real-time as possible. Video and image process models make a lot of sense here. Since there may be a lot of areas to cover, or the images may have some latency issues getting to the cloud, this can help improve your response time to quality issues.

- You don't want to send all your data to the cloud. This could be true if you have sub-second data streaming from your equipment. It may be possible that you do not want to send all this data to the cloud, incurring bandwidth, upload, processing, and storage costs, but if you process, visualize, and aggregate locally, then the value can be extracted without the cloud overhead. Aggregates or summaries can be sent to the cloud for additional analysis.

There are now many solutions to help you address some of the cases, AWS IoT GG being one of the offerings from AWS that provides a variety of edge capabilities. A sample GG architecture is presented in *Figure 6.8*. As with most AWS services, GG is a toolbox that enables you to deploy and manage applications and services you define to your edge servers running GG within the factory.

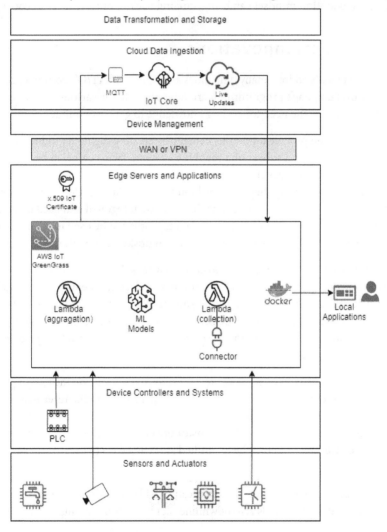

Figure 6.8 – Sample GG architecture

The possibilities are endless, allowing you to connect and collect data from various systems and process and forward data to your cloud applications. In addition, the integration AWS provides with the rest of your application architecture allows things to flow seamlessly and securely across the network. In later chapters, we will go much deeper into these topics, but now is an excellent time to think about your systems. Begin the categorization and detailed understanding of options within your environment. Think about the deep legacy equipment that requires the interpretation of custom serial protocols. We have all faced this challenge and require another level of sophistication within our integration efforts.

As we have seen, architecture decisions go a long way in determining the future state of an enterprise. Being agile and open to new and cutting-edge technologies is key. Let us explore the options and opportunities an open adoption mindset can bring, circling back to our digital twin concept.

Adopting technical innovations

New technology developments address many difficulties associated with IT/OT convergence. Low-code applications and natural language programming are now possible for various devices and sensors. This programming eliminates the need for coding knowledge to get them online.

Digital twins, as we noted, can provide numerous benefits by creating digital reproductions of objects. This can be a whole machine or a subset of a machine system. Twins can also be single pieces or assemblies that can be used to model changes to the system or allow considerations, such as stress testing, to mimic failure. Alternatively, they could be brand-new product developments modeled within a comprehensive approach to help determine how the parts are integrated and used more effectively. Inspection can be done using virtual and augmented reality, or it can be used to improve training or allow a trained professional in a remote location to assist an unskilled operator with repairs.

Let us look at the applicability and fitment proposition of a car's digital twin depicted in *Figure 6.9*. A taxi's digital twin is created by deploying IoT sensors across all components. Multiple stakeholders benefit from this data. The OEM of the car gets real-time feedback on the product's performance on the road, which directly feeds into future product enhancements. The OEM also gets geography-specific performance information, which can help design better-suited products in price-competitive and value-driven market segments.

The component supplier also gets an opportunity to continuously innovate on the design and optimize for performance and cost by analyzing real-time data. A suspension manufacturer may, for example, calibrate products based on local road conditions. The ride-share company can get ahead of the curve by cleverly analyzing driving patterns and fuel consumption, paving the way for better service for the end customer. The incentives can thus be quantified and passed on to the driver-partner without ambiguities.

Service and maintenance providers also benefit as they can provide near real-time assistance empowered by system-specific data and work towards zero downtime, significantly improving end-user satisfaction.

The insurance companies, at the other end of the spectrum, tremendously improve settlements and claims based on data and can detect any possibilities of fraudulent claims in real-time. The digital twin powered by IoT is a win-win proposition for all the stakeholders involved.

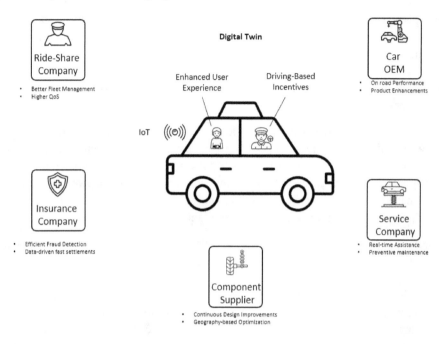

Figure 6.9 – A case study for a car digital twin

Finally, edge computing can be introduced where large data streams often cause bottlenecks in the cloud. With edge computing, data is fully or partially processed before being sent to the cloud, resulting in an analytics platform that may be better tasked for analytics, and minimized machine latency.

While many have started and are leading their digital transformation journeys, IT/OT integration is one of the fulcrum pivots. We will see how various industries are faring in this perspective.

Which industries are leading the way?

The industries most vulnerable to change are leading the way as IoT technologies evolve. In recent decades, key industries have received significant investment and previously imagined developments. The sectors listed here are excellent instances of successful IT/OT convergence.

Companies may now integrate the two processes thanks to IT/OT convergence. OT departments may strategize in real-time to accommodate changing demand, and IT can use inventory updates to inform price changes. Additionally, the greater cybersecurity that IT departments provide to OT can help all industries. While the closed structure of most OT systems has led many to feel immune to cyberattacks, this is usually a false sense of security. Due to a lack of authentication or encryption, unconnected systems may be more vulnerable. Bringing IT expertise into the OT domain can help firms in various industries become more secure.

Data availability is quickly transforming factory processes in the manufacturing business. Manufacturing plants are being transformed by IT/OT convergence, making existing procedures more efficient and cost-effective. In typical manufacturing systems, IT and OT are separated into production and demand cycles. OT systems collect data from the manufacturing floor to guide product cycles and inventory management. IT systems handle market shifts and client conversion rates, which generate predictive analytics for future demand.

Natural-resource processing and distribution, such as oil and gas, are increasingly reliant on digital networks. Geographical dispersion and complicated equipment made unified organizational functionality impractical in the past. Because OT infrastructure is dispersed across continents, local teams were historically responsible for manual operations, maintenance, and updates. However, technological advancements have allowed OT teams to remotely access operational data, improving how they undertake equipment inspections, damage assessments, and inventory management.

The energy sector has made major IT/OT convergence expenditures in recent decades. Collaboration between IT and OT in energy industries can help assure regulatory compliance and defend a company's brand. IT technologies also aid in the tracking of energy distribution and the synchronization of maintenance schedules. Electrical system maintenance may be expensive and time-consuming, and in today's 24/7 economy, equipment failure and unanticipated downtime can have significant financial ramifications. Engineering teams can plan preventative maintenance with access to machinery data, reducing risk and downtime.

Finally, in the transportation business, IT/OT convergence is helping to improve operations. Rail companies, as well as bus and metro operators, can benefit from integrating IT and OT systems to acquire a better understanding of asset conditions and usage. This can help with short-term repairs, asset replacement, and safety planning. As we see the emergence of e-mobility devices such as electric bikes, scooters, and buses, this convergence is becoming increasingly crucial in the transportation business. Experts predict that better decision-making, more straightforward maintenance, and safer, more reliable devices will result from coordinating the physical assets at the junction of IT and OT.

Summary

We have looked in more detail in this chapter at the challenges and opportunities of industrial integration and the digitization of your industrial solutions. We have taken a step back in technical detail to offer more insights to determine direction and decision-making. We have continued to explore architectural options for you to consider as you design and build your solution.

As we have seen throughout the various sections in this chapter, IT and OT convergence, although complex, is inevitable for a successful digital transformation of an enterprise. Developing a keen understanding of the complexities and devising solutions based on industry-leading open architecture patterns will pave the way for a future-proof convergence and seamless data flow northbound and southbound.

Besides the technical challenges of availability (or lack) of data, connectors, protocols, security measures, and so on, fundamental to convergence is the unified mindset of the stakeholders from the shop floor to the top floor. At the end of the day, as we re-emphasize, plants and processes without considering the people would be near useless in terms of usability and perceived design value.

In many industries, PLCs are at the heart of manufacturing processes. They control the machinery that drives the industry. In the next chapter, we will look in depth at the technical details of PLC integration and industrial protocols and review ways to collect and send that data to the cloud for processing and analysis.

## 7

PLC Data Acquisition and Analysis

This hands-on chapter will help readers understand and appreciate the integration of a **Programmable Logic Controller** (**PLC**) with a data acquisition system. Starting with the architecture and build of a PLC and moving on to the evolution of hardware from the 1960s, this chapter will shed light on the actual brain controlling the machines on a shop floor. We will also introduce the readers to the basics of programming a PLC with ladder logic and the tools for this from various **Original Equipment Manufacturers** (**OEMs**). The important aspects of configuring the protocols, mapping the tags, and retrieving the data from PLCs is handled by showing the readers how it's actually done. This is the fundamental integration point for OT and IT.

We will delve into the following topics in this chapter:

- PLC hardware and architecture
- PLC programming: ladder diagrams
- Practical application of PLC: smart mocktail bar
- Data integration and mapping
- Data ingestion and data analysis

Technical requirements

This chapter provides a good overview of industrial control systems. We will deconstruct a PLC to understand the inner workings of the automation device that single-handedly fueled Industry 3.0. The second half of the chapter is dedicated to advanced topics that drive the industry today and will tomorrow. The first half is straightforward, while a basic understanding of industrial networking would be a plus for the second part of the chapter. Knowledge of the following would be a great starting point to enjoy this chapter:

- Control systems
- Field bus protocols in the OT domain
- Basic networking concepts
- Security fundamentals

PLC hardware and architecture

PLCs mainly are built around the central processing unit, a real-time and deterministic processor. It is interfaced with memory units that store programs and data. Most PLCs run on 24 or 48 V DC supply. PLCs are also built modularly by having pluggable cards that can be expanded and customized based on the application scenario. The **Input/Output** (**I/O**) cards or modules are usually expandable and can be configured. These cards typically deal with analog and digital signals.

High-speed **Pulse-Width Modulation** (**PWM**) sampling circuits are also built in, making the PLCs capable of processing fast signals in real time. The communication modules are, last but not least, key here too. There are communication ports for programming the PLC, obtaining diagnostic information, and interfacing with peer PLCs, networks, and sensors. They help transmit the information upon which the external data logic can be built. *Figure 7.1* shows the hardware architecture of a generic PLC.

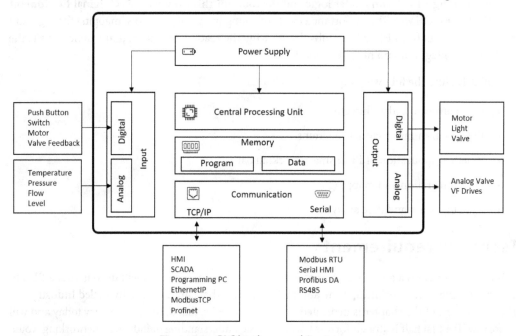

Figure 7.1 – PLC hardware architecture

The critical aspect is that most of the PLCs support serial and TCP/IP-based communication. Serial protocols such as RS485, MODBUS RTU, and PROFIBUS DA are supported for data transmission.

TCP/IP-based protocols provide higher throughput, communicating with **Supervisory Control and Data Acquisition (SCADA)** systems, **Human Machine Interfaces (HMIs)**, and peer PLCs in the network via EthernetIP easier. PLC architecture has more or less remained constant, with a few exceptions now in terms of communication, storage, and processing power.

> **PLC remote I/O**
>
> We have looked at the I/O modules attached to the PLCs. Their primary function is to convert data from sensors into a digital format that's readable and actionable by the PLC – and vice versa. Remote I/O comes into play when it is not possible to house the PLCs near the field devices (such as in very harsh environments). These remote I/Os are a cost-effective way to implement ruggedized control, as expensive controllers are housed at a distance while I/O units are stationed close to the field devices. The communication is then handled through a high-speed bus to maintain real-time control.

Let us take a look into understanding how PLCs operate.

Operating PLCs

PLCs execute via the process of scanning. The scanning frequency can range between 1 to 20 ms, translating into 1,000 to 5,000 times each second. Each PLC scan has 4 distinct components as represented in *Figure 7.2*.

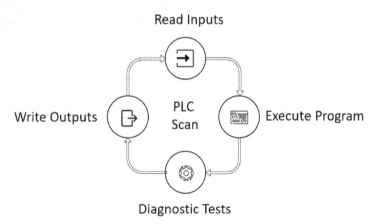

Figure 7.2 – PLC scan cycle

The PLC reads the inputs (analog or digital). The inputs are read only once per cycle and written to an image. The outputs are written once at the end of the process. Input signal changes during a cycle are ignored. This is important because, in the worst case, the signal runtime or processing time can correspond to twice the cycle time. The ladder logic or the program is then executed left to right from top to bottom (most OEMs) where the input data is used to perform logical and data-based operations. Then, the PLC performs diagnostics check to ascertain whether there are any memory errors. If an error is detected, the PLC stops execution and restores the outputs to the default normal state. This step is crucial, as manufacturing applications are mission-critical (machine and human). Finally, the outputs are set per the calculated executed logic. The scan is thus repeated over and over again in this manner.

To understand and appreciate the functioning and programming constructs of a PLC, it is imperative to dive into a few fundamentals of electrical systems and control systems. While ladder logic programming has evolved from electrical engineering circuits, control systems are used heavily to modify system behavior. *Figure 7.3* depicts a simple electrical circuit to control a lamp. The positive of the power supply is connected to a switch and the other end of the switch is linked to the load or bulb. The neutral is then wired to the other terminal of the bulb. When the switch is flipped, the circuit is completed, the line is energized, and power flows through the load, thus illuminating it. When the switch is open, the load is electrically isolated from the power source. This logic is important as we design ladder logic diagrams to program PLCs.

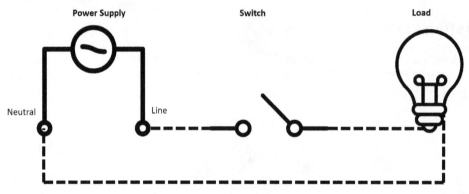

Figure 7.3 – PLC programming roots from electrical circuit diagrams

Process control systems comprise control loops as shown in *Figure 7.4*. This can be broken down into two major parts: the system and the controller.

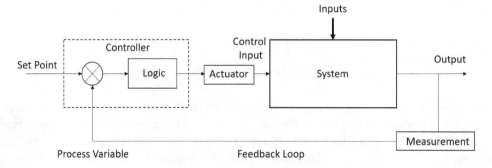

Figure 7.4 – Control systems fundamentals

The system has few inputs and behaves in specific ways to provide output, but to obtain the desired output, we need to create a feedback loop. Thus, the output is measured, and the process variable is fed back into the controller. The controller already has a set point or a standard operating value. The controller then applies built-in logic to produce a control input. This input is then actuated into the system to produce the desired output. The whole objective of the control system is to minimize the difference between the output measured and the desired set point. In the practical example in this chapter, we will look in detail at process variables and control inputs (also called manipulated variables).

With an overview of control systems and electrical engineering fundamentals, we are now ready to delve into PLC programming. Since ladder diagrams have a direct correlation with electrical relay diagrams, readers will be able to appreciate the correlation.

PLC programming – ladder diagrams

Let's build on our understanding from *Chapter 5, OT and Industrial Control Systems*. We have seen that ladder diagrams are the most popular frame of PLC programming adopted in the industry and they have evolved from electrical relay diagrams. The two vertical lines represent the power and ground, and each rung (horizontal line) represents the logic to complete the circuit conditionally. Contacts and coils form the two major components in each rung. A contact *usually* represents input while output is represented by a coil. If a path can be established from the left side of the rung through the asserted contacts to the right side, then the rung is true, and the output is asserted. If the path cannot be followed, then the actuator remains un-energized, thus deeming the rung false.

> **Important note**
> Contacts can also have outputs, which we will discover via an example in the last section of this chapter.

Let's take the following example to comprehend the constructs of a basic ladder diagram. There are two input switches (**Input 1** and **Input 2**). **Input 1** is a NO contact, which means it is in a **normally open** state. The **Input 2** switch, on the other hand, is an NC contact, implying that it's **normally closed** by default.

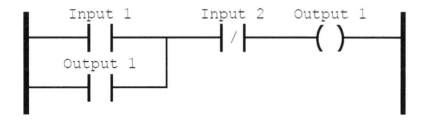

Figure 7.5 – PLC latching with contacts and coils

According to the preceding diagram, when the **Input 1** switch is closed, the **Output 1** coil is energized. Note that **Output 1** is LAO used as a contact. This is the concept of latching used in industrial applications. Most of the switches in the industry are push button-based and so have an inbuilt spring mechanism. Therefore, when the switch is released, it goes back to its default state. To maintain the same output even without holding the switch in an *on* state, the concept of latching is introduced. How do we break the latch, or in other words, how do we turn the output *off*? We use **Input 2**, a second switch that is closed by default. When the Input 2 switch is pressed, the circuit is opened and the latch is broken, deactivating the output coil. These fundamental concepts can be applied to all aspects of PLC programming.

Figure 7.6 represents a simple logic diagram for an embedded control system.

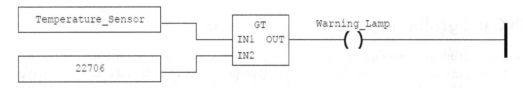

Figure 7.6 – Temperature control using function blocks

The program's objective is to switch on the warning lamp whenever the temperature exceeds a specific set point. There is a comparator block that takes in two input values. The first comes in from the temperature sensor and the second is the set point (**22706** is the equivalent of 40° Celsius according to the Steinhart-Hart equation). The real-time comparator energizes the warning lamp coil whenever the current temperature measurement exceeds the set point.

As discussed in *Chapter 5, OT and Industrial Control Systems*, IEC 61131-3 is the first independent and internally accepted standard for industrial automation languages. The most significant advantage is that it is independent of the underlying hardware, making seamless interoperability in a highly heterogeneous ecosystem a definite possibility.

Figure 7.7 depicts the IEC 6232-3 software architecture model in detail.

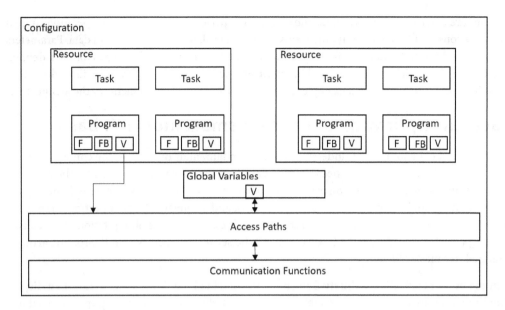

Figure 7.7 – IEC 61131-3 software architecture model

At the outermost layer, *configuration* represents the entire system or a control problem. A configuration entails hardware specifications, memory addresses for I/O channels, and system functionalities. Multiple resources can be defined within a configuration. Resources are processing facilities that execute or are responsible for the system functionality and operational logic.

Resources encapsulate *tasks* and *programs*. Tasks are responsible for one-to-one or one-to-many programmatical executions. Each execution is either time-driven or event-driven. Programs, as we saw earlier, can be constructed from either of the five approved programming languages. For the rest of the chapter, we will deal with ladder logic, the most used form and visually comprehendible form of PLC programming.

> **PLC secure programming best practices**
>
> We have stressed the importance of securing the OT layers and that security starts from the root – that is, from the PLC programs. Some of the best practices are as follows:
>
> - Employ PLC integrity checks using crypto hashes or checksums
> - Restrict and audit HMI input variables
> - Track PLC operating modes to identify any tampering
> - Lock PLC program access and unused communication ports
> - Modularize and document functions and error flags
> - Run PLCs in an isolated and segmented network (a VLAN) and only allow P2P connections
> - Implement network security for traffic analysis and possible threat detection

Programs are made of *functions* and *function blocks* that perform data and logical operations in real time. They consist of a data exchange mechanism and a data algorithm to transform data. Parameters can be defined through available data types with an option for derived data types for user-defined custom *variables*. Variables can be global in nature or local for each program. We will be dissecting this architecture using a practical example to appreciate the various components of the architecture.

Practical application of PLCs – a smart mocktail bar

Let us undertake the journey of understanding the fundamentals of PLCs in the context of smart manufacturing by considering a specialized assembly line for preparing mocktails. The mocktail filling station consists of a menu display, a QR code generator, five stations of varied juices (iced tea, lemonade, grapefruit, pineapple, and watermelon), and a conveyor belt connecting them. Each station has a QR code scanner and an infrared-based line detector. The filling stations each consist of a pump, a solenoid valve, and a flow sensor. All the sensors, actuators, and the robotic arm are connected to a master PLC.

From the user's perspective, a thirsty Bob enters the mocktail bar. He is welcomed with a display of multiple mouthwatering options such as pine lemonade, fruity mocktail, grapefruit iced tea, water(melon)-lemonade, lemon tea, pinemelon tea, and so on. This is depicted in the sample user experience diagram.

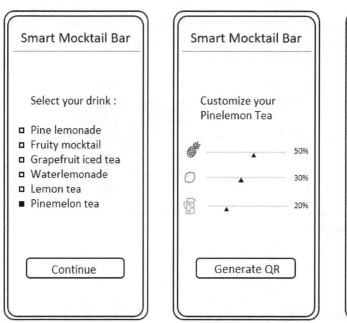

Figure 7.8 – Smart mocktail bar user experience

Bob can also control the percentages or ratio of individual ingredients per his taste. He chooses his favorite mocktail, pine-melon tea, with a stronger pineapple ratio. He then goes on to select the size of the cup and completes the payment from his virtual wallet. The QR code generator prints a custom QR code that the user pastes on his cup and places at the starting point of the filling station. As the cup travels to each station, the IR sensor stops the cup precisely to fill it. The QR code reader detects the ingredients list and passes the volume information to the controller. The controller dispenses the exact volume of the juice at each specific station. Finally, when the cup reaches the end of the line, it is stirred, and ice cubes are added. Now, Bob has his favorite custom mocktail ready to be slurped. The entire application landscape is shown in the following diagram.

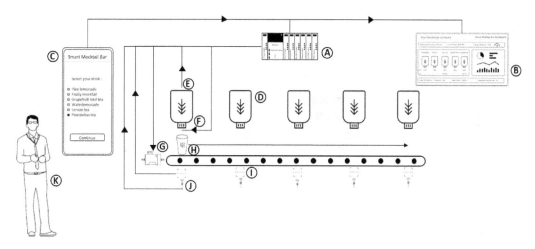

Figure 7.9 – Smart mocktail bar setup

A. PLC or the central controller

B. Edge store dashboard

C. User input screen

D. Juice dispenser

E. Level sensor

F. Solenoid valve (dispense control)

G. Conveyor belt motor

H. Mocktail cup with QR code

I. QR code scanner

J. IR line sensor

K. User ordering a mocktail

The system consists of various sections and inputs. At the highest level, it can be broken down into five filling stations consisting of various types of juices. Each of them is equipped with a level sensor (to indicate when it goes below a certain threshold) and a flow sensor to measure the flow rate based on the material's viscosity. Each filling station also has a QR code scanner that instructs the PLC on the cup's presence and the juice's respective composition in the given filling station as per the user's choice. In essence, each filling station has five sensor inputs feeding the PLC. The PLC controls two outputs: the conveyor belt motors to stop and start the movement of the cup and the solenoid valve of each filling station to control the inlet of the juice into the cup. This I/O–simplified model is presented in *Figure 7.10*.

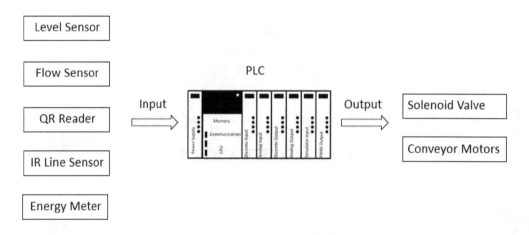

Figure 7.10 – PLC sensor inputs and output

The representation is simplified, of course, for easier understanding. The next section describes the energy meter interface via the Modbus protocol in detail. The PLC is at the heart of the system, providing precisely controlled operations in real time. It takes in inputs from multiple sensors, such as an IR line sensor, an optical reader, a flow meter, and so on, and actuates output, such as a solenoid valve, pump, conveyor stepper motor, and more.

Looking at this use case from a PLC software architecture perspective, the configuration consists of the I/O addresses to acquire data from sensors and control actuators. The resources are then defined for the application based on the hardware design. Since each filling station performs a similar operation, we can create a global function block that can be called every time the action needs to be invoked. To intrinsically calculate line efficiency, global variables can be defined to measure common parameters and **Key Performance Indicators (KPIs)**.

The ladder logic is presented in *Figure 7.11*. As discussed in the initial section, the ladder logic consists of contacts, coils, timers, and flags. Multiple vertical rungs with system logic are implemented to cover corner cases. Once the user input for the mocktail is received, and the user places the cup at the starting point, the conveyor motor is energized. The IR line detector placed at each filling station detects the arrival of the cup at its location and the conveyor motor is turned off. Now, the QR code reader is activated and the product composition data from the current filling station is extracted. Then, the quantity of juice required from the current filling station is deciphered as a variable of time, and the solenoid valve motor gets activated. After the requested quantity is dispensed (the counter runs down), the dispenser solenoid motor is turned off. The flow sensor is also employed to collect actual feedback. This process is repeated for all five filling stations until the cup is filled as per the initial user's requirement. The ladder diagram is abstracted to give the user a basic understanding of the logic. Additional logging functions and custom function blocks are employed for data collection and standardization.

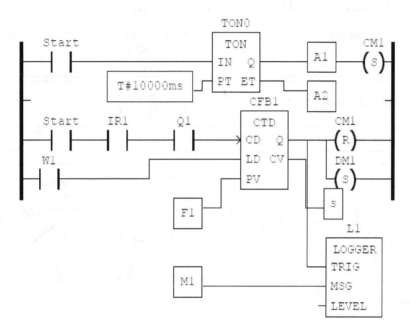

Figure 7.11 – PLC ladder diagram

Although this example looks like a classic case of a process manufacturing filling plant, it has a lot of built-in nuances to make it a smart mocktail bar. To start with, the product at hand is custom prepared as per the user's input in a mass production, optimized, streamlined unit. This is a great shift from our traditional manufacturing, adapting to changing user needs.

Next, intelligence is embedded into the product. While the machines have control and intelligence, this is a step further where the product decides and informs what add-ons or customizations are needed for a given station.

Distributing intelligence to the product gives way to many use cases and applications. The mocktail use case can also be extended to the end of life where the user's feedback on taste can be intrinsically captured from where they dispose of the used cup. Product on-field performance feedback and end-of-life feedback translate into the likelihood of acceptance by the end user. These are very important for filed KPIs that can influence the entire product design, marketing strategies, etc. There could be two bins to capture the feedback, a *like* bin and a *meh* bin. Depending on how they liked the drink, the user can throw their cup in either of the bins by scanning the cup's QR code. Depending on this data collected over a period of time and from all users, we can refine the product next time by suggesting best-fit products based on user demographics, choice preferences, geographical region, and statistics such as best sellers. Supply chain visibility, tracking, and auditing are also built into the product and can be leveraged at any point and place to trace back origins and details such as timing, production line details, batch numbers, and more. This is a pivotal step when we think of big data applications such as mass product recall, value-added production decisions, etc.

The innovative mocktail bar dashboard displayed in *Figure 7.12* provides a real-time view of the production facility. It has data flowing in from the order board and the PLC.

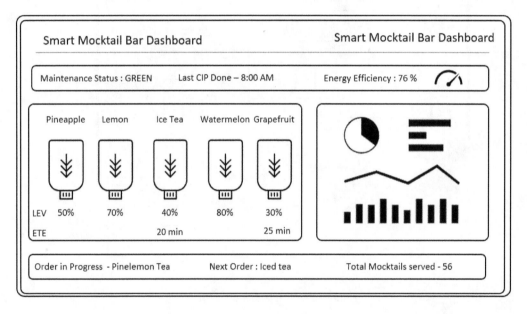

Figure 7.12 – Smart mocktail bar dashboard

While showing the order in progress and the orders in the queue, the dashboard has pivotal information regarding the operations of the smart mocktail bar. These inputs are consumed by line operators and maintenance technicians to schedule product refills, cleaning in place (planned maintenance), and so on with ease and accuracy to minimize downtime, thus improving operational efficiency and user experience. Downtime reduction through timely refill notifications, maintenance during non-peak utilization based on historical data, and automatic cleaning in place are some of the clear winning use cases for enabling data-driven Industry 4.0 use cases.

One interesting use case worth mentioning is smart rerouting. Let us assume that the smart mocktail production is in progress and the cup has reached a station. The filling station is currently being refilled since it is empty. Now, the PLC can reroute the cup to the next filling station and bring it back to the first station when it's filled and ready for operation. Not only does this optimize time, but it also makes decisions based on real-time changing requests, adapting itself for the optimal outcome.

The cloud connectivity on top of this provides even more firepower. While being a reservoir for practically limitless data storage and compute, multiple potential applications surface. For example, a real-time comparison of various mocktail bars worldwide is a great use case that is now made possible. Historical data analysis, running predictive machine learning algorithms, asset performance management, and energy-based production optimization are all advanced topics we have dealt with throughout this book. The possibilities created by obtaining data from a humble PLC controlling outputs based on inputs and scheduled logic now provide a lot of potential.

The fundamental aspect we have learned throughout the book lies in the holy grail of IT/OT integration. As we shall see in the next section, data integration and global aggregation play a significant role in application and operations optimization.

Data integration and mapping

PLCs are the powerhouses of automation and can independently operate and execute programs based on program logic – so, why do we need to aggregate and visualize data? Data monitoring and handling of data are very important to understand the current state and debug and diagnose in case of failures. The human or operator will continue to play a very important role in the entire process. To enable this human or operator to take important decisions based on human-comprehensible information, data acquisition and visualization become important. The PLC program, once finalized, is locked into the PLC and real-time ladder diagram debugging becomes strenuous.

The first step of data visualization is the HMI module at the production site. The PLC sends real-time data to the HMI, and production status with alarms and error codes is notified to the operator. The HMI also provides the user with the ability to alter predefined variables in accordance with the current process. For example, the packaging line needs to be set for the total planned production for the day and other important parameters such as products per box and the bases on which the machines will be tuned. The HMI is tightly integrated with the PLC and allows for direct read/write access to the PLC data variables. The HMI is usually at an individual machine or production line level where it has direct access to the addresses and the variables of the PLC.

SCADA systems are positioned at a plant level, where they integrate with multiple PLCs, HMIs, and so on. They provide a singular plane of reference for the plant manager to visualize and control production on the shop floor. Since SCADA systems operate with heterogeneous PLC builds and OEMs, there needs to be an abstraction for accessing and modifying PLC-specific data. This concept is called PLC tagging, where variables and addresses for a specific PLC are mapped to a specific PLC tag. This tag is exposed to the SCADA system for visualization and manipulation. Historians support systems for logging historical operational data over time to help visualize trends and derive improvement areas.

Data acquisition, as such, presents a familiar dilemma. Depending on the timing, the target persona, the context, the nature, the importance, and the actions associated with data are entirely different.

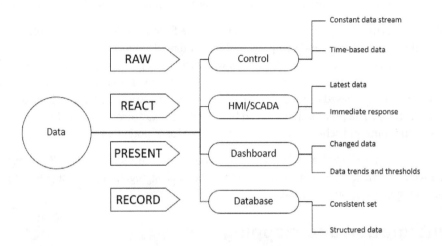

Figure 7.13 – The data acquisition dilemma

Data can be classified into four buckets, as visualized in *Figure 7.13*:

- For control purposes, data needs to be time based, real time, and consistent at the lowest level. This is raw data with a very high sampling rate. PLCs use this data to calculate formulae and logic and respond with real-time actuation.

- The next sample of data is presented to the human operator—these alarms and event data from outliers of the time-based constant data stream. The data from this funnel needs immediate attention to continue with harmonized production.

- The next filter of data is the dashboards, where further sampling shows trends, comparisons, aggregation, and KPIs over time. Here, changes and thresholds are presented at a macro level to make decisions at a holistic level.

- The last bucket consists of structured storage of data for data science-based operations. Trending and predictive performance optimizations are all possible with this data. A comparison of production data with financial and planning data year on year is a good example.

While these four buckets present a data dilemma, the sampling, labeling, and aggregation of data are application- and architecture-specific. Data labeling and mapping thus play a critical role in assigning importance to data and channelizing it through the different streams to obtain varied end results. While the overall objective is operational excellence, cost reduction, and so on, the means and techniques to get here differ.

The last section of this chapter is dedicated to discussions on data acquisition and analyzing the acquired data to take corrective actions. This has two interesting use cases: one talks about energy meter integration with PLC and the other circles back to **Overall Equipment Effectiveness (OEE)** calculation and analysis. The energy-based production operation is an excellent segue to sustainability topics, such as carbon footprint and ecological efficiency. They can be very vital parameters to compute, compare, and correlate production costs from an energy perspective. Since this slightly deviates from the book's focus, we will conclude the topic here. Nevertheless, it's a very valuable topic in today's context, and readers are encouraged to delve deeper into this subject.

Data ingestion and data analysis

An interesting use case in the production facilities is to interface energy meters to the central PLC. Traditionally, energy meters are part of utility management and are isolated from the production facility. They are present at an aggregated level and are usually used for metering purposes, but interfacing energy meters to production-line PLCs opens up a plethora of opportunities: line-wise energy and cost comparison, shift-wise energy utilization (used for time-based tiered energy costs), demand planning, and so on. We will focus our efforts on the process of integration, namely using the Modbus protocol.

Modbus is one of the oldest yet simplest field bus protocols based on master-slave topology over a serial line, RS232/RS485. The protocol was introduced for communication between PLCs and other systems. *Figure 7.14* shows the interface of the energy meter to the PLC using the Modbus protocol.

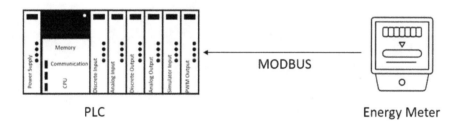

Figure 7.14 – PLC and energy meter integration using the Modbus protocol

The master initiates the communication request with a function code and a data request. The slaves receive and respond with the corresponding data. The simple communication architecture is represented in *Figure 7.15*.

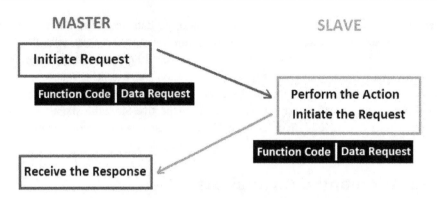

Figure 7.15 – Modbus master-slave communication topology

While many modern energy meters support Wi-Fi, LoRaWAN, and so on, most products still adopt Modbus as a communication protocol. To initiate communication between the energy meter and the PLC, some parameters, such as device address, baud rate, and parity bit, need to be configured in the energy meter UI panel. Depending on the manufacturer of the energy meter, different power parameter readings are stored in unique addresses (registers). The device register map is the single most important data sheet that is needed to map the correct data registers to the requested energy parameters. This data must be programmed in the PLC to query for a particular parameter at a given time. Parameters such as current, voltage, power factor, power, energy, and error codes can be transmitted to the PLC from the energy meter as depicted in *Figure 7.16*.

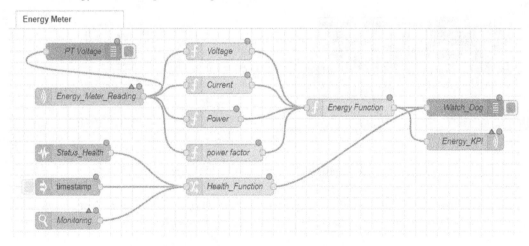

Figure 7.16 – Node Red flow for energy meter data integration

After looking at energy management integration, we will return to operational data management using PLCs. The analogy in this section will help us connect the dots from the initial parts of the chapter. *Figure 7.17* is a diagram restructured from an earlier example of our sample production line. This has the addition of the PLC and the HMI for controlling an injection molding production line. The idea is to measure the operational efficiency of the production line from the data obtained through the PLC.

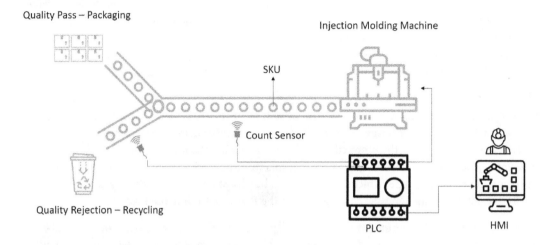

Figure 7.17 – Operations management using a PLC

The data ingestion and transmission from the PLC to the cloud are discussed in detail in *Chapter 11, Remote Monitoring Challenges*. This section, however, considers local calculation and display of KPIs limited to the shop floor. OEE is a measure of how well equipment lines are utilized in relation to their full potential. We looked briefly at the calculation of the OEE in *Chapter 3, In-Situ Environment Monitoring*. Here, however, we will only be looking into the significance of OEE as an operation metric and ponder over other meaningful KPIs.

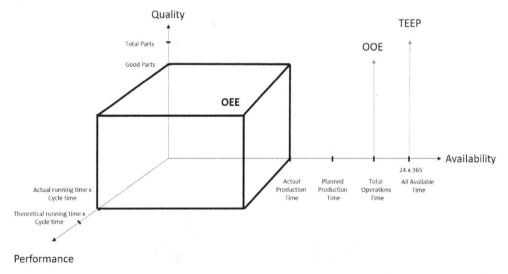

Figure 7.18 – OEE, OOE, and TEEP

OEE is the product of three percentages: availability, performance, and quality. It can also be visualized as the cuboid volume formed in *Figure 7.18*, which depicts availability across the *x*-axis. All available time, that is, 24 (h) x 365 (days) of operational time (the theoretical limit), is plotted at the extreme end of the *x*-axis. Considering the total time, we arrive at a KPI called **Total Effective Equipment Performance (TEEP)**. This parameter gives the maximum performance of a line or a machine when there is no downtime from bad-quality products. TEEP is used for system design for capacity benchmarking. This can also be used as a go-to golden performance standard for targeted optimization initiatives.

The next plot is of the **Total Operations Time (TOT)**. This does not include unscheduled downtime (such as machine failures, operator emergencies, and so on). The KPI calculated by employing the TOT as the time quotient is termed **Overall Operations Effectiveness (OOE)**. When the calculated OEE inches closer to OOE, the production facility has fewer unscheduled downtimes. This is a great milestone for the operations and maintenance teams working in close collaboration. Next, we have the planned production time, which includes lunch breaks, shift changeover times, and so on, and finally, the actual production time coming in from the PLCs controlling the machines.

Along the *y*-axis, we have the quality metric plotted, giving a measure of total parts produced and total good parts produced. The *y*-axis has the plots for the performance of the asset. Here, the term cycle time is used, which defines the effective time to produce a good part. Data analysis can be employed here to evolve the cycle time based on actual data, historical averages, and product-dependent metrics.

OEE is calculated at an SKU level, the lowest attributable element in the manufacturing ecosystem. It can then be calculated at a machine level, aggregating to a production line, a shop floor, and an enterprise. OEE is thus a significant metric for continuous improvements in the operations of a manufacturing facility.

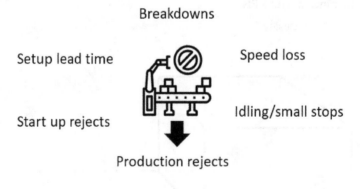

Figure 7.19 – Equipment production losses

There are six major categories of equipment production losses, as shown in the preceding figure. Some are attributed to process and maintenance schedules, while others depend on operator efficiency, knowledge, etc. OEE gives instant and real-time feedback on the type of loss, helping quantify steps for continuous improvement.

Summary

This chapter started with a deep dive into PLCs, the brains behind industrial production and operations. While PLCs have been a very closed confidential space with many OEMs using proprietary techniques and implementation architecture, we endeavored to present a standard hardware construct and software architecture. Moving into the realm of PLC programming, we chose ladder diagrams owing to their simplicity, popularity, and industry adoption rates. Starting from the electrical analogy and the basic constructs, we dived deeper through a production line control application.

The smart mocktail bar was an example application providing users with a glimpse of production automation. The application was packed with multiple fascinating Industry 4.0 use cases, aligning it with the central theme of this book. Next, we introduced the data tagging and mapping concept leading toward northbound data integration.

The next chapter is an interesting one, focusing on asset performance management. A key topic that is most important in an operational context, we will delve into the details through practical runbook-style execution using AWS services. You may now turn the page!

8

Asset and Condition Monitoring

Improving the efficiency of your systems by reducing cost and time or improving the quality of your manufactured product is a primary goal of Industrial IoT. Unfortunately, it takes time to achieve these goals, to determine what to measure, and to understand how to measure and react within your systems and processes. Even if you deeply understand your running production systems, when adding new measurements to the mix, you need to correlate the data and incrementally progress your understanding of how to react to various changes in this data.

As stated, one of the primary goals of Industrial IoT is to improve the operational efficiency of your equipment. This is not an overnight process to achieve results with instrumentation and monitoring improvements; instead, it is a progression and a maturity curve that needs to be adopted and evolve as you move forward. Many organizations live within an initial area of **reactive** or possibly **preventative** maintenance, depending on the type of **enterprise asset management (EAM)** system or maintenance procedures they currently have in place.

As we initialize data and start collecting and analyzing more from individual assets, we can move toward condition-based maintenance. We can look closer at the real-time conditions around that asset. Measurements are compared to help us understand if a component is underperforming and may need to be looked at more closely. At this point, we look toward not doing maintenance too early or too late but in line with actual feedback from the asset and its performance.

Moving to predictive or prescriptive maintenance can take this further by adding **machine learning (ML)** capability to your data analysis. This allows for faster and more accurate system monitoring based on what you have seen in the past or what criteria you might decide to look at, removing the human component from the monitoring process and providing that feedback based on what has been learned about optimum operating conditions.

In this chapter, we're going to cover the following main topics:

- Understanding **asset performance management (APM)** and system monitoring

- Investigating data processing architecture
- Making meaningful decisions with your data

Technical requirements

We will learn the need for measuring real-time asset conditions and move toward a better understanding of intelligent monitoring. This chapter is more technical than some of the others. An excellent working knowledge of **Amazon Web Services** (**AWS**) IoT will be helpful, although we will try to explain as much as possible along the way. Some basic programming is discussed, but we feel the examples are relatively simple.

You can find the code samples mentioned in this chapter at `https://github.com/ PacktPublishing/Industrial-IoT-for-Architects-and-Engineers/tree/ main/chapter08`.

Understanding APM and system monitoring

We titled this chapter around APM which is the ability to monitor and assess the performance of your equipment and processes across the factory floor. Unhealthy assets can contribute to a wide array of issues across the manufacturing process, from unexpected downtime to reduced productivity and output quality. Organizations can measure asset performance, availability, and quality using **Overall Equipment Effectiveness** (**OEE**) metrics; however, knowing what is going on at the equipment level takes a deeper approach to understanding. Any specific asset within your manufacturing environment contributes to your factory's overall efficiency and productivity. The goal is to enhance efficiency by monitoring, maintaining, and possibly improving each piece of equipment within the chain.

We examined *Figure 8.1* early in *Chapter 1* of this book, *Welcome to the Revolution*, and looked at how we can incrementally progress maintenance initiatives for our production lines and equipment:

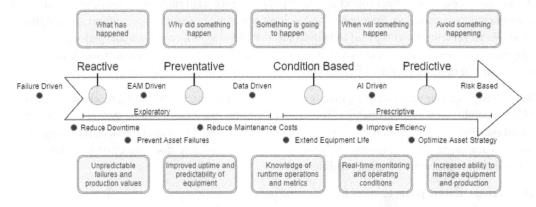

Figure 8.1 – Maintenance and monitoring roadmap

In addition to maintenance, if something is broken, there is another level of interest that can help us understand if things are working correctly. This can have several connotations. First, it can help us know if our equipment is working at peak efficiency. Fluctuations in temperature, vibration, pressure, and even subtle changes can affect the speed or quality of output at scale. Second, this type of monitoring can provide additional detail into potential problems in the future, essentially using ML to evaluate and look for potential issues.

What is APM?

Asset performance can be a confusing topic. There are several definitions as vendors try to define a vision for their customers. Overall, we see APM as monitoring and managing the performance and availability of our assets and machines. APM 4.0 uses several approaches to achieve this goal. *Figure 8.2* shows the relationship between APM and other monitoring and management techniques, such as condition-based monitoring, which we consider a subset of overall APM:

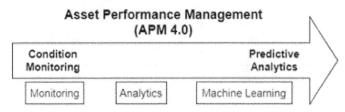

Figure 8.2 – APM overview

In this view, **condition-based monitoring (CBM)** is a subset of the overall APM area of focus. There are slightly different viewpoints on these topics in the industry, but whether you agree with this alignment or not, or maybe not entirely, we can agree in principle that our roadmap and goals are the same. To move further toward the goal of Industry 4.0 adoption, we must use a common approach to monitoring, analyzing, and applying ML to our equipment data streams.

APM at scale

This chapter primarily focuses on data collection, processing, enhancement, and some initial analysis. But to be clear, these are all steps that need to be done in every case. As with everything complex in nature, you have to crawl before you walk, and monitoring and analysis are no different. We must understand the domain and the equipment and define our measurement and analysis goals. This is especially true when thinking about doing anything at scale. Understanding the data, your analysis, and what you are looking for in the data is essential. Implementing ML without this basic information won't get us very far. We can start by slowly walking through our data to understand more about our data and how we might look for patterns that can help us know when our assets and equipment are in trouble or possibly be noticeably improved. The following section, focusing on data processing architecture, follows precisely this process: providing you with a starting point.

Investigating data processing architecture

We mentioned this earlier in this book: we love it when operators start seeing their equipment in new ways. Operators can usually see some parameters about the equipment, especially with more current operations, but they often want to see more data differently. This is where IT and OT can work together. IT can show the equipment and processes as data, whereas OT can share how the physical equipment and processes work. Domain knowledge and data analysis go hand in hand in this case—the result is exponential improvements in trust, and the partnership grows.

As usual, before we get too deep into solutions, we want to look ahead and see where we are going. In this case, we will leverage some of the same AWS services considered earlier but more in-depth to flush out a few additional pieces of the architecture. *Figure 8.3* illustrates our data processing path and goals for this effort. We will take each layer and outline our specific focus for that area:

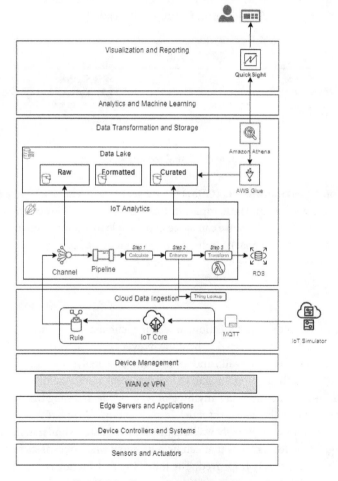

Figure 8.3 – Data processing architecture

There are only three main layers in play for this effort, so we will skip the others and focus on those three:

- **Cloud data ingestion**: We have seen cloud ingestion before, so there is not much new here, but we are adding a new simulation system to help us build a prototype and test our efforts moving forward.

- **Data transformation and storage**: This is the heaviest section, with many moving parts. In *Chapter 2*, *Anatomy of an IoT Architecture*, we skimmed over AWS IoT Analytics, but here we embrace it entirely and outline how to use it for maximum effect.

- **Visualization and reporting**: In this layer, we open up our **business intelligence** (**BI**) capabilities with Amazon **QuickSight** to provide a simple interface for some analysis to help understand the data in more detail and provide helpful information and insight.

We will start by using simulated data for this chapter. We will look at how we can request data from **programmable logic controllers** (**PLCs**) and other devices in a later chapter. Mock data is a great way to separate work happening at the edge and on the factory floor for your development purposes. Data must be processed and analyzed efficiently. Using simulated data, you can easily update proposed data formats or introduce conditions that would be harder to replicate with live equipment. Simulated data is best used to build your infrastructure and develop basic flow. Once you have things in place, you can turn to historical or even live data to monitor the conditions you expect within the system.

Simulating data for processing

In years past, we built simulators using Java or Python to send sample messages at regular intervals. The concept is well known and can be accomplished with only a few lines of code; however, sending anything other than simple messages at regular intervals complicates things. It gets messy trying to do anything more complex in most cases, like trying to simulate a fault or conditions leading up to a problem. In addition, there is a case of security when using AWS IoT Core. Another necessary consideration is to generate and use certificates within your code base to ensure the data is transmitted correctly.

Fortunately, sending IoT data messages has become more of a commodity service, and there are powerful tools that can alleviate the stress of building your simulators. MIMIC MQTT Simulator is a robust tool that allows for complex situations. The MIMIC Simulator Suite has a set of products to help you in your Industrial IoT journey.

AWS IoT Simulator (linked next) is another tool that can provide a framework and engine to send data of whatever type and interval you require: `https://aws.amazon.com/solutions/implementations/iot-device-simulator/`.

We will work solely with the AWS simulator here. The setup of the simulator is not complex but requires a working knowledge of CloudFormation and knowing how to define the necessary parameters correctly. It may take a couple of tries, even if you are familiar with running CloudFormation templates, but if not, you may need to request some additional expertise in getting it up and running.

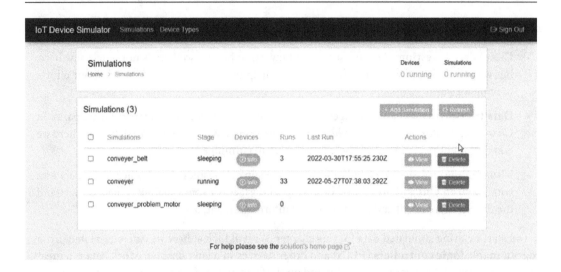

Figure 8.4 – IoT Device Simulator

With **IoT Device Simulator**, you can set up complex devices with multiple measurements and then combine those devices into simulations. In our case, we created a simple device representing a conveyor belt motor. We started a pool of four `conveyer_motor` instances, each with a different serial number (`sn`) with measurements of random values within a specific range. These values are defined when we set up our device with various values. The resulting measurements are illustrated in the following snippet:

```
{
    "vibration": 100,
    "rpm": 3250,
    "voltage_phase1": 390,
    "voltage_phase2": 410,
    "voltage_phase3": 385,
    "sn": "361774893",
    "datetime": "2018-02-15T12:50:18",
    "timestamp": "1518699018000",
    "current": 290,
    "temp": 150
}
```

Note that the voltage is provided for each phase of our three-phase motor. We will use that later to determine if the motor is supplied enough voltage or too much based on its rating. `sn` (the serial number) is used to identify each motor and to look up metadata around the device. Admittedly, this is a simple example of what we should measure, but it provides an example of how we might get started with the process. Start with what can be measured, which data is available, what can be added, and how we can make sense of the data.

> **Important note**
> We included both a `DateTime` value and a timestamp in the message because we were not sure how we would process the message initially. IoT Analytics can use both types of values in most cases, although the timestamp does need some transformation later in this chapter. So, one of these may end up not being used, and we can eliminate it in the actual messages. This is one of those trial-and-error moments where you are unsure what should be used, so you test with both.

For this case, we focus on the voltage level across three phases, with an approach that if a voltage is too high or too low, we could have a problem over time with our motors. Too high means the motor runs hot and burns out quicker; too low and the motor will struggle to perform at its rated capacity. Ideally, we want all three phases to be in sync and within an acceptable percentage of rated voltage between each other.

We set our simulated data to run for an hour at a time and send a message every few seconds. That could be considered a lot of data if it were running 24/7; however, this is not an issue in our case. Do consider cost and evaluate when testing with many hundreds or thousands of messages. Too much velocity and things could get expensive quickly.

Consider the interval of data capture and processing effort in the real world, or move this entire analysis to the edge for quicker real-time alerting. One more thing to note is that the simulator is set up to send data to IoT Core automatically; hence the incoming messages arrive on whichever topic you define when you create the mock device within the simulator. From there, it is a matter of defining a rule within AWS IoT Core that can forward your messages wherever you need them for processing. In our case, this will be an IoT Analytics channel that we will see in more detail in the following sections. As messages arrive, you can configure the channel directly to pick up data from an IoT Core MQTT topic. Either approach will work, but be consistent across your channels and overall strategy.

Now that we have our initial message defined and can send messages to IoT Core, we will focus on how to process messages once they arrive.

Processing data with IoT Analytics

Chapter 4, Real-World Environmental Monitoring, briefly discussed AWS IoT Analytics as an option for processing incoming data in a streamlined manner. With everything AWS, there are a lot of good choices and a lot of flexibility in how you want to handle things. Sometimes this can be frustrating as you wish AWS could just decide for us, especially when you embark on a new topic or technical area. Still, this flexibility allows teams to design the right system for any particular use. There is strength in this flexibility—or complexity if you are new to an AWS service.

At its essence, AWS IoT Analytics is a simple concept of a channel, where data goes in to start, and then a follow-on pipeline or set of pipelines, where this data can be processed in any way you can imagine. There are a few additional components, which we will call out as we go:

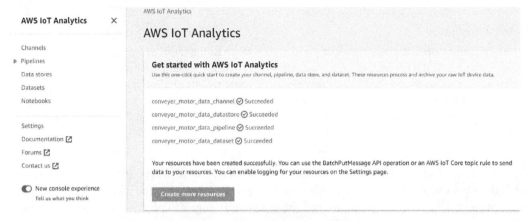

Figure 8.5 – AWS IoT Analytics: wizard setup

The preceding screenshot, *Figure 8.5*, shows a set of components created automatically based on the available one-click quick-start option. In this case, we are setting up a set of components based on our `conveyer_motor` example. The wizard will set up a channel, a pipeline, and a data store for use in the process.

Entering the channel

Getting data into and initially processed by an IoT Analytics channel is almost magically easy compared to many new technologies. *Figure 8.6* shows the setup for our channel to process `conveyer_motor` messages:

Figure 8.6 – IoT Analytics channel

Channels do a fundamental job; after a little configuration, they can perform essential tasks without any intervention, such as the following:

- Receive or accept incoming MQTT messages from a specific topic
- Store those messages into **Simple Storage Service** (**S3**) via a configured data store as raw data messages
- Forward messages to preconfigured IoT Analytics pipelines

That's it! There is not a lot to think about at this point. Your data goes to the S3 location you have defined. For me, that is this location: `S3 bucket / Channel / <channel name>`. You can configure your data store with a little more information, such as using custom dimensions and defining which format you want to store your data. As of this writing, either JSON or Parquet formats are available.

Note about data stores

I didn't go into much detail about configuring data stores in this chapter. They require an S3 bucket and a role configured to access that bucket. I created a generic role called `IoTAnalytics_ChannelServiceRoleForS3`, which had the required permissions to access the raw data lake S3 bucket.

So, your raw data is stored. Congratulations! This significant first step provides comfort, knowing that your original data is safe and sound. It would be best if you aimed to get this data ingested and stored as efficiently as possible. This allows you to spend processing time and effort elsewhere to make this data more consumable. You may notice that a channel does not allow you to modify messages. This is by design since the integrity of the data may, at some point, come into question. Ensuring that the data remains unprocessed helps you defend its accuracy or truthfulness.

The other side of that coin is retrieving the data at some point when you need to track the provenance of the data higher in the value chain, if necessary. For our example, we checked that the data was being stored and moved on. Still, additional effort may be required for large-scale data volumes to ensure that your folder structure and dimensions are clear enough to assist in looking up data in the future when needed. It's also necessary to note that the payload messages are not decoded for any data type. That comes later and will need to be considered in your own processing effort.

Finally, the raw data storage should be fairly locked down. It should be immutable, which means it cannot change or be overwritten. Once you are sure data is being stored the way you like, put these permission blocks in place as soon as possible to protect your information's reputation as a **source of truth** (**SOT**). Additionally, you may look in on the data from time to time and add some archival rules to move older data into cold storage or delete it if necessary.

Managing the data pipeline

The next step is using a pipeline to transform the raw data into something useful. That process depends significantly upon the data and your goals. Initially, there may be some decoding of payloads or adding additional fields to messages.

At the core, a pipeline is precisely what it says. It takes your message and moves it along a path, making changes along the way. We could have multiple pipelines that perform different actions or enhancements, carrying data from raw to formatted (or cooked) and then even further transformed into enhanced data ready for ML applications or advanced analytics:

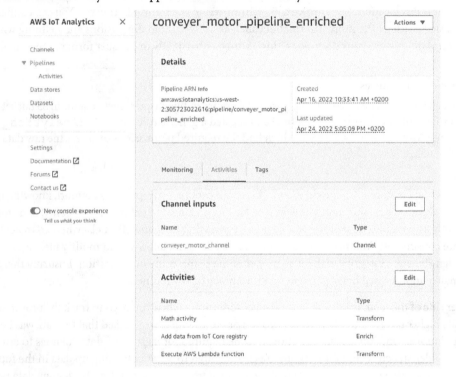

Figure 8.7 – IoT Analytics Pipeline

Figure 8.7 shows the enriched data pipeline that defines multiple activities in processing messages—specifically, to perform the following activities within the pipeline:

1. Perform a simple **math activity** converting the temperature from °C to °F.

2. **Enrich** our data from attributes stored in IoT Core that tell us more about the motors we monitor.

3. Execute a **lambda** that performs several functions, such as cleaning up the JSON format, adding data from an external database, and converting our timestamp to something usable.

We could do these activities in many ways, especially since we have a lambda function in our midst; however, using a managed service as the core of guiding our actions is comforting and allows us to change or update the pipeline with minimal effort. We can even rerun data through the pipeline after we make some changes. Let's start with a simple math activity to see how simple changes to our data can be accomplished.

Creating a simple math activity

If all you need to do is perform simple calculations on your data, AWS IoT Analytics is the best thing since sliced bread. No lambdas, nothing tricky; define a simple math calculation and create new attributes to represent those values easily:

Figure 8.8 – Pipeline math activity

In *Figure 8.8*, we perform a simple conversion calculation to convert the temperature from °C to °F. It is a bit contrived as an example but perfectly illustrates how this feature can work. Another example may be to convert an analog value to its actual value. Many analog sensors output a voltage (for example, 2.4 mA) corresponding to the true value of what was being measured. For example, an analog temperature sensor outputs its value along a range of voltages, which needs to be converted to the measured temperature. The pipeline allows you to remove attributes from the message when configuring the system. Adding new attributes is also a single-step process. The pipeline activity provides an additional view of seeing your data before and after the calculation, so you know what you are getting.

Admittedly, this is one of the simplest ways to transform your message and one of the least powerful, but it's excellent for what it was designed to do, which is to perform simple equations on your measurements. Let us look at some more complex pipeline transformation and enrichment options.

Enriching a message from an AWS IoT thing

IoT thing types are familiar objects when defining devices within IoT Core. Things help define a physical object and allow you to identify the object, adding attributes, descriptions, and configuration data associated with a thing type. In this example, we have defined a thing type called `Conveyer_Motor`, along with multiple instances of this type. Note that the name of our thing in *Figure 8.9* is the serial number of the actual motor it represents. This allows us to match up the motor later to ensure we can track this particular piece of equipment:

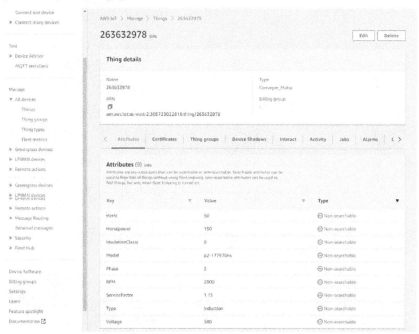

Figure 8.9 – AWS IoT thing attributes

In this case, we have added a set of attributes specific to our motor, allowing us to determine the rated values of the equipment. The rated voltage, horsepower, RPMs, and insulation class will help later with some analysis. Admittedly, not every message needs this information, resulting in some bloat during storage and processing. Still, it's an excellent example to start with and helps you understand some of the options available. Maybe you want to do additional math calculations based on this data in the next activity?

There is an architectural decision to be made here. Do you leverage the attributes and information about your devices or things? Or do you use a custom database to store information about your machines? We will do both in this chapter so that you can see how each works and make some design decisions. At scale, we are going to need to think about interfaces. How do we update the attributes and description data about devices we deploy or replace in the field?

Figure 8.10 shows how we can enrich a message during pipeline processing. Again, IoT Analytics makes this pretty simple. We are using the message's serial number (sn) to match against IoT Core things and pull in all the attributes associated with that thing. The activity setup page shows you the message before and after the enrichment, so it is easy to see how your message will look at each stage in the pipeline. Since this activity takes place after the math activity, notice that the tempF we created earlier is now part of the ongoing message:

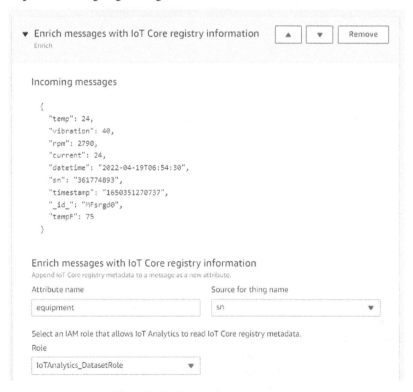

Figure 8.10 – Data message enrichment

After processing the updated information, the attributes are added to the message as a nested object within the JSON structure. There are also some attributes we probably don't need or want, clogging up storage and processing, so we need to think about this structure. *Figure 8.11* shows the enriched message data that has been added from IoT Core Thing attributes:

```
Outgoing messages
Below are the attributes to be included in the outgoing message.

{
    "temp": 24,
    "vibration": 40,
    "rpm": 2790,
    "current": 24,
    "datetime": "2022-04-19T06:54:30",
    "sn": "361774893",
    "timestamp": "1650351270737",
    "_id_": "MFsrgd0",
    "equipment": {
      "defaultClientId": "361774893",
      "thingName": "361774893",
      "thingId": "ae248b9b-8ce6-418f-b20a-62b9dfd2dd45",
      "thingArn": "arn:aws:iot:us-west-2:305723022616:thing/361774893",
      "thingTypeName": "Conveyer_Motor",
      "attributes": {
        "Type": "Induction",
        "Phase": "3",
        "ServiceFactor": "1.15",
        "Horsepower": "150",
        "Voltage": "460",
        "Model": "p2-177978nx",
        "InsulationClass": "B",
        "Current": "278",
        "Hertz": "50",
        "RPM": "2900"
      },
      "version": 5,
      "billingGroupName": null
    }
}

Update preview
```

Figure 8.11 – Enriched message

As a future discussion point, this type of nested structure will probably not work well with our analytics or BI needs. Complex JSON structures often need strong interpretation by BI and analytics systems. The data may come across as an opaque nested object during cataloging. There are ways to ensure this data gets identified and cataloged, but we want to keep our message structure relatively clean for analysis. We will flatten this structure and make a few additional changes before later storing this within the curated data store. *Chapter 12* will examine how to convert data to a CSV file for data modeling as another example.

Using a lambda function as a pipeline activity

For our last and most complex activity, we will trigger a lambda function to perform several transformations to the message. In essence, our lambda function will perform the following modifications:

- Flatten the JSON structure to remove nested objects

- Pull in some simple data from a relational database

- Modify the timestamp to fix formatting discrepancies so that the data store can use it for dimensions

We will talk a little more about the lambda function structure during this section to show you some ways to be sure you are designing your functions correctly. While we do not specifically focus on application architecture in this book, the development team must have a straightforward approach to coding standards and conventions, as discussed in *Chapter 2, Anatomy of an IoT Architecture*. Let's add the transformation Lambda function to operate on our data:

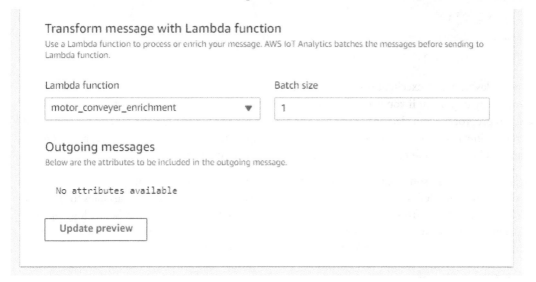

Figure 8.12 – Lambda transformation

We have set up a lambda function called `motor_conveyer_enrichment` and identified this lambda in *Figure 8.12*. The batch size has been set to 1 to keep things simple. You may want to increase the batch size, especially if you have a lot of incoming messages and want to transform them more efficiently or bundle them together. Our lambda is not designed to work with a large array of messages in the incoming event JSON, so one at a time is initially sufficient for our needs.

> **Important note**
>
> Generally, you can preview a message both before and after any activity to see your message as it is transformed. For a lambda transformation, previewing the transformed message within this view is not possible since it doesn't have insight into the returned message format. We have removed the initial message from *Figure 8.12* for brevity, but it is the same message that is defined as the outgoing message in *Figure 8.11*.

The lambda function is not complex; however, several actions are taking place within the process. Let's start at the beginning and look at the `import` statements defined for the function. Specifically, we will identify the inclusion of the `pymysql` (Python MySQL) library. This is not a standard library within lambda Python, so extra effort was taken to include this library as a lambda layer:

```
###Define Imports
import json, os, sys
import time
import logging
import pymysql
```

Lambda layers are an excellent way to include libraries or dependencies within your code base. This is especially important if you want to use libraries that are not readily available—for example, if you want to preprocess your data for ML uses. Adding a layer is not extremely difficult, but it is also not trivial for inexperienced developers. Once you do it, it becomes pretty straightforward, and there are many available web references to help you install specific lambda layers.

Next, we will set up some parameters within our lambda function. This is all prep work to ensure all our settings are defined in a single location. Note that we use environment variables to store the parameters separately from our code. In the following section of code, we describe the logging level to define which messages get logged to CloudWatch, and then set up the variables needed to connect to the database using our Python MySQL library we imported earlier:

```
###Setup logging level
logger = logging.getLogger()
logger.setLevel(os.getenv('loglevel'))

###Setup connection to MySQL
DBhost = os.getenv("hostURL")
DBusername = os.getenv('username')
DBpassword = os.getenv('password')
DBname = os.getenv('database')
charSet = "utf8mb4"
cursorType = pymysql.cursors.DictCursor
```

Environment variables can be a powerful feature in separating external values and variables from your code base. For example, if you had a lambda function running in your development environment and in production, the information required to connect to the necessary database could be more configurable:

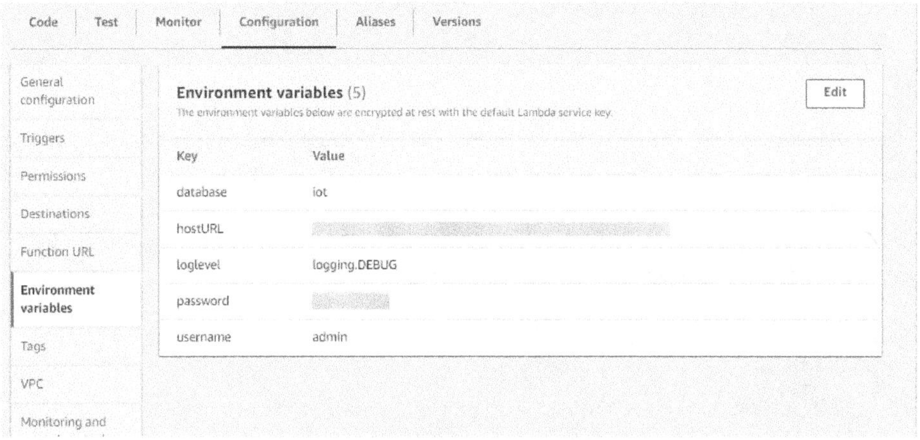

Figure 8.13 – Lambda environment variables

Figure 8.13 illustrates some of the environment variables we use in this lambda function. Next, we enter the `lambda_handler` function and dump the incoming event data. This is strictly for identification as we test the lambda to be sure we are starting with the right input:

```
def lambda_handler(event, context):
    #logger.debug(os.environ)
    logger.debug("event: " + json.dumps(event, indent=2))
```

We mentioned how the IoT Core Thing attributes are added to the message as a nested object within the JSON structure. This can make things messy later in the process, so we want to spend a little time flattening out the structure. We are identifying the attributes we want and making duplicates at the top level of the JSON. Later, we will drop the originals and the entire nested object when we have what we need from the incoming structure. There are many ways of flattening a JSON structure; since our needs are simple, we do it manually. But for complex structures, a library would be helpful. Here's the code we need to execute:

```
    ###flatten some interesting attributes for use in analysis
    event[0]["rated_insulationclass"] = event[0]['equipment']
['attributes'].get('InsulationClass', None)
    event[0]["rated_servicefactor"] = event[0]["equipment"]
["attributes"].get("ServiceFactor", None)
    event[0]["rated_rpm"] = event[0]["equipment"]
["attributes"].get("RPM", None)
```

```
    event[0]["rated_voltage"] = event[0]["equipment"]
["attributes"].get("Voltage", None)
    event[0]["rated_horsepower"] = event[0]["equipment"]
["attributes"].get("Horsepower", None)
    event[0]["rated_current"] = event[0]["equipment"]
["attributes"].get("Current", None)
```

Next, we need to make a simple change to the timestamp value. The IoT simulator provides a timestamp in milliseconds, which is quite common. For the data store to use the timestamp dimensions for storage (that is, year, month, day, hour), we need to convert the timestamp to seconds rather than milliseconds. This is a simple issue with how IoT Analytics interprets the value. To accomplish this change, we will divide the timestamp by 1000 and replace the existing value:

```
    ###convent timestamp to integer for datastore to use as
partition
    #print(int(time.time())) #current unix timestamp
    #event[0]["timestamp"] = int(time.time())
    event[0]["timestamp"] = int(int(event[0].get("timestamp",
None) ) / 1000)
```

I am using an integer instead of a double for simplicity. You may need more precision by performing `double time_seconds = time_milliseconds / 1000.0;`.

Next, we want a simple database lookup to enhance our message further. We know the insulation or temperature class from the thing attributes. We want to perform a simple database lookup to determine the rated temperature for that class. We extract the temperature class from the message, defaulting to class B if nothing matches. Then, we create a simple query to look up the class in our table:

```
    ###Get temperature max from insulation class attribute
    insulationClass = event[0]["equipment"]["attributes"].
get("InsulationClass", 'B')
    query = "SELECT temp FROM iot.temperature_class_lookup
WHERE class = '" + insulationClass + "'"
    logger.debug(query)
```

Figure 8.14 shows a simple lookup table called `temperature_class_lookup`. We designed and manually created a MySQL database in Amazon **Relational Database Service** (**RDS**) to hold our lookup table, which could be the start of something much more significant when dealing with the metadata of our sensors. Experience with **database management systems** (**DBMS**) and some data management expertise is a fundamental skill for most experienced architects, so we do not detail how to build and populate this table. Honestly, a lot of metadata is usually required for organizational assets. Much of this can usually be found in enterprise systems such as asset management systems:

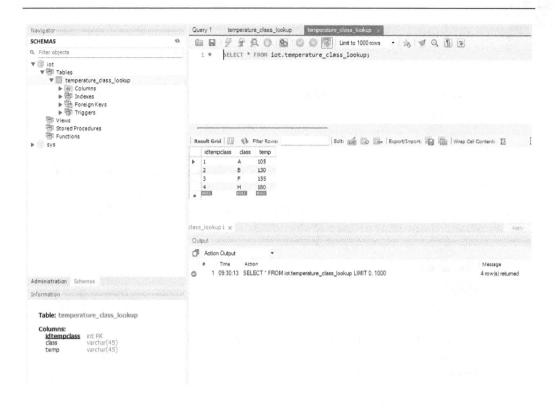

Figure 8.14 – Asset database

An EAM system generally holds essential data about your machines, serial numbers, maintenance records, and when components were placed in service or replaced. We shouldn't need much of that data here except to enhance our specific messages or perform in-place calculations that might be needed later. But the concept of a lookup is the same. Enhancing data in flight through a simple table or an organization's extensive system is similar. The following set of code provides the lookup to enhance our message:

```
try:
    # Create a connection to the Database
    logger.debug("Connecting to DB")
    conn = pymysql.connect(host=DBhost, user=DBusername,
passwd=DBpassword, db=DBname, connect_timeout=5,
charset=charSet, cursorclass=cursorType)
    logger.debug("SUCCESS: Connection to RDS MySQL")
    # Create a cursor object
    cursor = conn.cursor()
    # SQL query string
```

```
            sqlQuery = query

            # Execute the sqlQuery
            cursor.execute(sqlQuery)
            #Fetch all the rows
            rows = cursor.fetchall()
            for row in rows:
                logger.info("rated temperature for this device = "
+ row["temp"])
                event[0]["rated_temp"] = int(row["temp"])
        except Exception as err:
            logger.error(err)
            #sys.exit()
        finally:
            conn.close()
```

I will make a short plea here to think about scale. Querying a table directly for hundreds or even thousands of messages per minute can start to be constrained. We would consider a service for caching somewhere along the line for data lookups such as these since it is repeatedly the same query over and over.

We have all the data we need for now in the message. At this point, we can drop the nested object that was initially provided by the earlier enrichment action. And then, for debugging and testing, we need to print out the final message to ensure the final message is structured correctly:

```
    event[0].pop("equipment")
    ###Output the result and return to caller
    logger.debug("event: " + json.dumps(event, indent=2))
    return event
```

OK—that was quite a bit to follow. But let's have a look at the finished product. This is the JSON after being processed by the lambda function. It has undergone quite a transformation in the process:

```
{
"vibration":37,
"rpm":3179,
"voltage_phase1":387,
"voltage_phase2":397,
"voltage_phase3":376,
"sn":"361774893",
```

```
"datetime":"2022-04-28T05:55:19",
"timestamp":1651125319,
"current":274,
"temp":43,
"tempF":109,
"rated_insulationclass":"B",
"rated_servicefactor":"1.15",
"rated_rpm":"2900",
"rated_voltage":"460",
"rated_horsepower":"150",
"rated_current":"278",
"rated_temp":130
}
```

This is a pretty good start at an enhanced data message. We want to consider more measurements, but with a strong pipeline in place, we can add to this as new sensors or edge integrations are completed. We can start some analysis and try to begin to discover how we want to use the data as it evolves. This also helps IT understand the domain better as they work with OT to determine which measurements make sense and which anomalies they should consider.

> **Note on message reprocessing**
>
> One more thing I should mention: IoT Analytics allows you to reprocess messages if you make changes. This is fantastic, especially when you are just starting. If you enhance your pipeline, you can reprocess the pipeline's data and update your curated data on the fly.

At this point, we can start collecting actual data. Assuming something changes, we know all the pieces to configure and adapt. Collecting data from real systems can be eye-opening, even for the operators. It allows them to understand what normal looks like with their equipment, and they never want to go back to when they were without it. Be warned!

Cataloging and querying curated messages

This section will look at the processed data and understand how we might use it effectively. We looked in earlier chapters at how we could send and store data in a time-series database for easy charting with Grafana. We can extend this approach to send data to relational databases or large data warehousing applications to be crunched in many different ways:

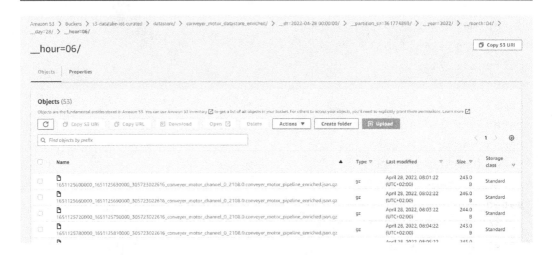

Figure 8.15 – Curated data in S3

Figure 8.15 shows the data stored in S3 after processing. This is now curated data for our example and is ready for deeper analysis, ML, or to be combined with data from external systems. The messages are stored in a well-defined folder structure that we will use as dimensions to help us target the data we need when we query.

> **ISA 95 and the Unified Namespace**
>
> We should point out that our folder or partitioning structure is relatively simple and specific to one set of equipment. As you look across your enterprise, you should examine a more unified naming convention for your time series data. This more extensive structure allows you to structure data across plants and zones within a plant, providing a much more detailed design for your data.

The files are stored compressed, with each file containing a single data record. Ideally, we would compress multiple records, perhaps grouping them by the minute or some other interval and using a data structure more appropriate for real-time analysis. Our chosen data format is JSON; however, if we batch these files into larger groups, we probably want to move to Parquet format, which is a columnar format, allowing for a faster query of the data files:

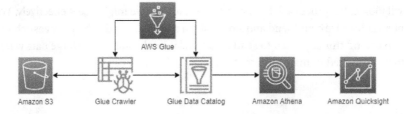

Figure 8.16 – Data cataloging approach

Moving further in our process, we will tie together several AWS services. *Figure 8.16* illustrates the services we will use for this, and we will discuss each step in the process individually.

Crawling and cataloging data in S3

AWS Glue provides a handy function. Our messages are in flat files in a folder structure (S3). This is not a helpful format for determining and especially querying the data within those messages. AWS Glue helps us overcome that barrier by rummaging through the folders we define and figuring out what is buried within:

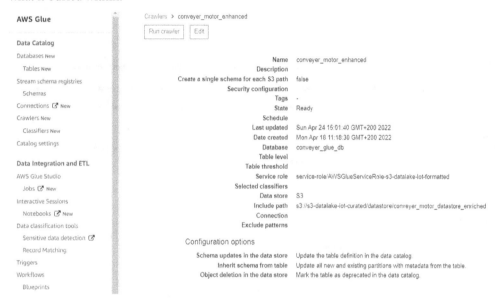

Figure 8.17 – AWS Glue crawler setup

Figure 8.17 sets up our Glue data crawler, which is quite simple. Review the folders and files we have identified and build a database with the results. Notice that we use the top folder as the starting path. This is because we want to include all the subfolders as identifying dimensions to our data, allowing us to query based on the serial number, or perhaps the UID, and different aspects of the date on which the data was collected and processed.

The information identified is stored in the Glue Data Catalog, and we have set up this configuration to update the definition if it detects changes in the structure of the messages or folder hierarchy. We have not defined a specific classifier for this effort; however, Glue has built-in classifiers for JSON and several data sources and types. We have already simplified our JSON structure, so the crawler will have no problem classifying and identifying the data structure. We can create custom classifiers for more complex data structures with nested objects and arrays to help us identify and structure the classification effort. *Figure 8.18* shows the final result of the crawling and classification effort:

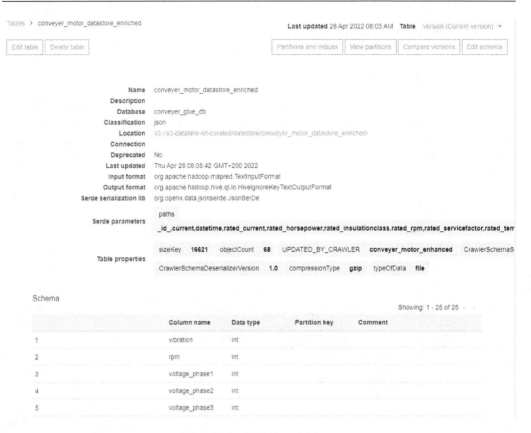

Figure 8.18 – AWS Glue database

AWS Glue has defined a table within our database with the same name as the crawler, which is fine for our efforts. The structure should look pretty familiar at this point, and Glue has done a great job identifying all our data fields and added a set of partitions based on the folder structure. Note that earlier in *Figure 8.17* there is a whole section in the menu focused on **Extract Transform and Load** (**ETL**). These features can be another way to transform, move, and enhance data from one location to another for additional analysis or availability to other stakeholders.

Querying data with Amazon Athena

AWS Glue helps us identify and catalog our data and imposes a structure on top of our data elements. Now, we can query our data and learn more about our systems with **Amazon Athena**. Athena is only a step toward using **Amazon QuickSight** to look more visually at the information being processed. Since QuickSight can use Athena to query the data that AWS Glue classified, we should look at how Athena works. *Figure 8.19* shows the Amazon Athena service interface with data, tables, queries, and results:

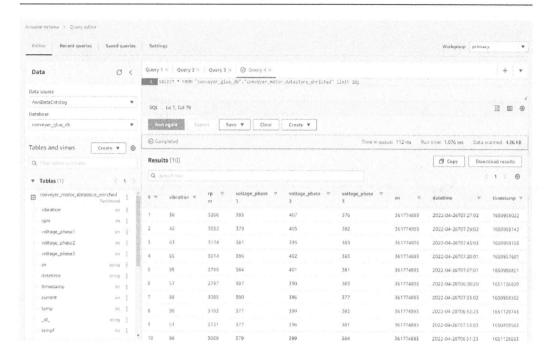

Figure 8.19 – Amazon Athena query

Amazon Athena allows us to query the Glue Data Catalog and view the data as a standard row/column result. It connects easily to the data catalog and uses standard SQL to construct a query and display the results. From here, we could do several things. Suppose this is your first look at the data. In that case, conducting some initial analysis of the data itself may be interesting to identify any potential issues with the data or the processing performed.

If the data looks clean for the most part, the data can be provided to other users for additional analysis and adding some BI capability. Amazon QuickSight works well with Athena and is designed to allow for BI from data across all your enterprise systems and data sources and is the next step in our journey.

Making meaningful decisions with your data

Finally, we are ready to visualize and perhaps make decisions or adjustments based on our previous efforts. You might've realized that it wasn't that much work, considering the gains to be made based on those efforts. It can be challenging to start from scratch or define a strategy for your entire enterprise. But this approach provides a great way to get your hands dirty and allows you to think about strategies you might use across your full architecture stack. End-to-end examples always give you insight into how things might flow, be stored, and, of course, how data should be best structured for visualization and analysis.

Amazon QuickSight is a BI tool. Learning a new and complex product, such as a BI tool, can take time and effort. BI software allows you to perform analysis and reporting on existing data. Still, there are nuances in how each product works and some of the core concepts of building a helpful report. One advantage of QuickSight is cost and availability; since your data is already stored within AWS, there is seamless integration, as you have seen throughout this chapter. Just remember—it will take some time and effort to get the hang of it. Amazon QuickSight is a powerful tool that allows you to provide data from many sources to your business users and analysts. Some in-memory storage capabilities use **Super-fast Parallel In-memory Calculation Engine (SPICE)**, which enables you to pull in data and provide fast access for more reactive reporting:

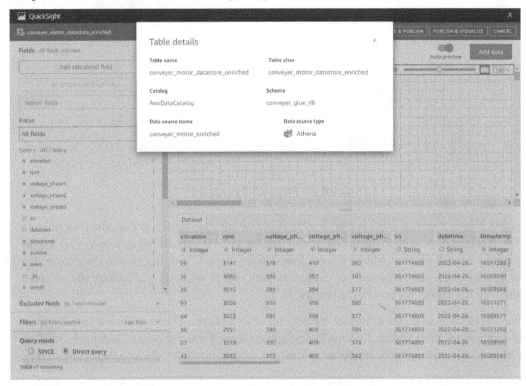

Figure 8.20 – Amazon QuickSight dataset

Figure 8.20 shows the creation of a dataset in QuickSight based on Athena as our data source. Remember—Athena allows us to query data cataloged using AWS Glue. It's a little confusing since the services are tied together but refer back to the set of services in *Figure 8.16* for reference on how they connect.

When you launch QuickSight, the easiest way to get started is to create a new analysis. An Analysis consists of a dataset and a visualization, although several steps help you set things up. *Figure 8.20* shows what a dataset looks like once it is set up based on using Athena from our example. As you review your data sample, realize that QuickSight allows you to identify the type of data for each column—that is, integer, DateTime, or string—or to modify or remove columns from your data source. Identifying the data type for each column is essential, so ensure you do this for your datasets as they are defined. It will help QuickSight later when analyzing or graphing data elements.

The next step in creating our visualizations is essentially drag and drop, but moving your data fields into the correct *X* and *Y* zone for visual analysis. Out of the box, QuickSight has several dozen graph and insight templates that allow you to analyze your imported data quickly:

Figure 8.21 – QuickSight visualization and analysis

Figure 8.21 shows some sample visualizations and insights created based on our data. We can see that the actual voltages across all three phases are well below our machine's rated voltage. However, it also appears that all three phases of voltage are relatively in sync with one another, which means that one phase is not spiking with regard to the other two phases. We can also see that the motor's temperature is within the rated temperature and does not seem to be spiking above the rated temperature for long periods.

Consider the monitor and evaluation needs of a single motor. We are initially looking at voltage and temperature here, but we also have RPM, horsepower, and current to consider. If the RPM or load on the motor ramps up, how does that affect temperature? Does the voltage increase with additional load or move beyond the rated voltage, which may also affect temperature, increase wear, and decrease the lifetime of the equipment?

We will dig a little deeper into some analysis and things we might look for as we move toward ML and how to use ML in your IoT systems. The visualization shown in *Figure 8.21* is helpful and allows several things to occur. We are first visualizing possibly new equipment or an operator has become concerned about it for some reason. The ability to visualize, analyze, and hopefully pinpoint issues is necessary. Secondly, it helps to use this real-time information toward driving your ML scenarios.

Summary

Architecting a good solution can be a challenging process. It includes understanding what is available and especially the outcome you need to achieve. We are still only scratching the surface in this chapter but hopefully, continuing to show you end-to-end solutions can help you make better decisions. Our samples enable you to mimic your prototypes or can be adapted to your specific solution domain.

This book section focuses on processing data from the industrial environment through a greenfield approach or adapting a brownfield location to provide the information you need to achieve and digitalize your environment. Along the way, we have focused your attention on real-world integration and data processing, allowing you to understand how everything can fit together without overwhelming the reader with too much detail.

In the next section, we will take things from an architectural perspective and look at some exciting technologies, such as edge computing, large-scale data processing, and incorporating ML into our processes. Again, the goal is to keep things understandable and allow you to follow the steps and emulate or adapt them to get something working that helps you move forward. With that in mind, let's get going.

Part 3: Building Scalable, Robust, and Secure Solutions

A well-built solution is not only a working prototype but should also focus on holistic aspects of architectural principles. This section will look at the essentials for designing IIoT architectures, encompassing the knowledge we gained from previous chapters. We will also examine how architecture will address intelligence at the edge and how machine learning works in an integrated environment.

As with most of this book, the chapters in this section weave between theory and practical hands-on examples. Each chapter builds on the previous set of materials. While you will understand how things work within each chapter independently, we will reference work in previous chapters that you can build on.

This part of the book comprises the following chapters:

- *Chapter 9, Taking It Up a Notch – Scalable, Robust, and Secure Architectures*
- *Chapter 10, Intelligent Systems at the Edge*
- *Chapter 11, Remote Monitoring Challenges*
- *Chapter 12, Advanced Analytics and Machine Learning*

9

Taking It Up a Notch – Scalable, Robust, and Secure Architectures

Strong architecture is fundamental to building an IoT system. This chapter will address the need for an architectural framework to develop scalable, secure, and robust **Industrial Internet of Things (IIOT)** applications. We will cover the broad spectrum of industry-wide architecture IIoT design considerations with references to the **Industrial Internet Consortium (IIC)** and **Reference Architecture Model Industrie (RAMI)** 4.0.

We start by building on the five S principles of sound architecture design and then look at system considerations. These are guard rails that every architect must consider as they design enterprise-scale industrial solutions. We will cover two critical tenets of architecture design for Industry 4.0: the **Asset Administrative Shell (AAS)** and the **Digital Service Bus (DSB)**. These concepts enable data interconnectivity and interoperability, thus creating value chains across verticals and horizontals.

We will also talk about the latest buzzwords – **Machine Learning (ML)** and **Artificial Intelligence (AI)** and the convergence with IoT enabling intelligence on industrial machines. Through an end-to-end manufacturing use case, we will validate the use of architectural frameworks and the value they bring in designing and commissioning an application.

We will delve into the following topics in this chapter:

- Understanding IIoT architectural principles
- Exploring IIOT reference architectures and how to use them
- Getting to know the AAS
- Getting to know the DSB
- System considerations for architectural design
- Leveraging architecture for AI and ML

Technical requirements

The chapter's objective is to build on the fundamental tenets of architectural design for the user. It aims to provide the user with a guiding framework as they endeavor to create both greenfield and brownfield enterprise IoT solutions. A basic understanding of architectural principles is essential to build on architectural design's various layers and dimensions. While this is a theoretical chapter, it sets the tone for creating solutions. Having said this, an understanding of the following constructs would help:

- Model-based design
- Design patterns

Understanding IIoT architectural principles

The Federal Ministry of Economic Affairs and Energy in Germany, in its vision for 2030, proposed three main pillars for shaping digital ecosystems globally: autonomy, interoperability, and sustainability. We can derive some key principles of architectural design from them. *Autonomy* focuses on the independent yet informed decision-making capabilities of individual components of the system. *Interoperability* helps create flexible networking between various stakeholders to create agile value networks. This has enabled standards and regulatory frameworks that aid in the seamless interconnection and exchange of information. *Sustainability* is a key pillar, focusing on a responsible manufacturing environment with social and corporate governance in mind.

The fundamental architectural principles or tenets are described in *Figure 9.1*. Simplicity, serviceability, standardization, scalability, and security – the five S principles for IIoT architectural design. There are other essential tenets, but they can be accommodated as subsets of one of these central pillars.

Figure 9.1 – IIoT architectural principles

Let's look at each of these tenets in a bit more detail:

- **Simplicity**: The most sophisticated and complex problems have simple and elegant solutions. The focus for a good architecture design should be to use minimal resources (storage, network, compute, hardware, software, and services) to solve a particular use case. This does not imply that architecture is localized to one application, not scalable and extensible in the future. It means that interfaces are modularized and not left open-ended.

- **Security**: "Secure by design" is today's enterprise architecture mantra. A robust and secure design is a must in a world of OT/IT integration, where critical assets run 24/7 to produce goods. Appropriate controls have to be in place from the ground up, and the philosophy of **Zero-Trust Any Access (ZTAA)** needs to be thoroughly applied. Access needs to be provisioned with the least privilege policy, and constant monitoring of threats needs to be carried out. With this, some observability has to be built into the architectural framework.

- **Serviceability**: Serviceability leans towards interoperability and openness of the system to interface with other systems. The exchange of data is critical to the success of the value stream as we will see through practical examples in this chapter. This also means that the solution can be packaged as a service and be replicable in a future site in a plug-and-play manner. Rather than the mere exchange of data between components, the focus is on cooperation between them in an automated and homogenized way. We shall focus on the openness to communicate further in the chapter via the DSB.

- **Scalability**: The motto of Industry 4.0 solutions has always been to think big, start small, and scale fast. True to this philosophy, architecture should have the ability to scale. In an era of fluctuating demand and changing consumer behavior, the manufacturing value chain should be designed to scale up and down. This complex topic involves optimal resource utilization to obtain higher efficiencies.

- **Standardization**: As we have understood throughout the book, the single most challenging factor in adopting Industry 4.0 solutions is a lack of standards. Parallel evolution of protocols and localized standards gets in the way of global expansion and system interconnection. A significant amount of effort is being spent to create channels to make data flow rather than use the data to create path-breaking analytical applications. The AAS we cover in this chapter is a revolutionary step toward standardization.

Now that we have elaborated on the **Non-Functional Requirements (NFRs)** for architectural design let us jump into two leading reference architectures. Architecture is a very complex topic with no one correct solution. Hence, we can often adopt a framework or guiding system to design a robust architecture for our problems.

Exploring IIoT reference architectures and how to use them

The reference architecture can be considered a framework of processes, methods, and tools that aid application building. It provides an abstract starting point based on which specific systems, solutions, and applications evolve. In this section, we would like to present two of the most prominent reference architecture patterns: the **Industrial Internet Reference Architecture** (**IIRA**) from the IIOT consortium and RAMI 4.0.

The IIRA

The IIRA is an architecture framework that encompasses the information, identifies the fundamental architecture constructs, and specifies concerns, stakeholders, viewpoints, models, rules, and conditions of applicability. Viewpoints can be defined as conventions framing one or more concerns. Concerns are any topic of interest about the system while stakeholders are persons or organizations with an interest in or concern for the viewpoint and the system. *Figure 9.2* shows the three-dimensional structure of the IIRA architecture framework.

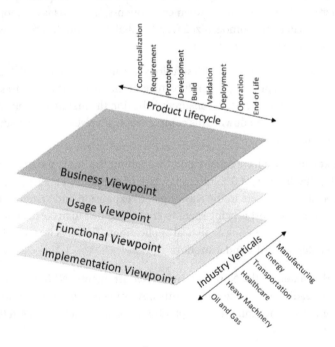

Figure 9.2 – IIRA architectural framework

The various viewpoints are abstractly presented as verticals. While we give the foundations of each viewpoint, we will dive deeper into the functional aspects, which form the crux of the system. We also have the **Product Life Cycle** horizontal axis. This tracks the progression of the production from the idea up until recycling. Along the other horizontal axis, we have **Industry Verticals**. The viewpoints and product life cycle may differ for each sector.

The different viewpoints, according to the stakeholders, are depicted as horizontal abstract layers. Business and regulatory concerns, such as business value, expected ROI, cost of maintenance, and product liability, are considered part of the given **Business Viewpoint. Usage Viewpoint** is an architectural plane that tackles how to implement the identified **Business Viewpoint**. It consists of various activities, tasks, and workflows coordinating work units over different system components.

The given **Functional Viewpoint** frames the concerns related to functional capabilities. The functional domain can be further broken down into five subdomains: **Control, Operations, Information, Application**, and **Business**. *Figure 9.3* represents the functional domain, crosscutting functions, and system characteristics.

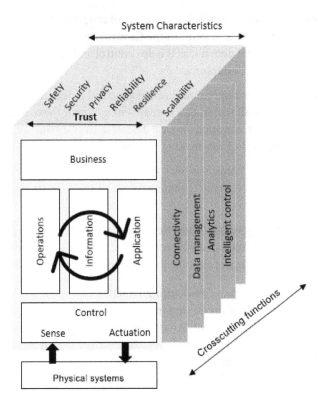

Figure 9.3 – Manufacturing types

The data and control flows between these domains are essential as we design a functional architecture. The functional domain will be described in detail in the following section. Finally, the given **Implementation Viewpoint** describes the technical components, their interfaces, communication protocols, behaviors, and properties of the system.

Diving deep into the functional viewpoint

The functional components focus on major system functions and are required to support IIoT use cases. This viewpoint aims to effectively actuate physical systems by processing the data captured by them.

In *Figure 9.3*, we see five major domains. These five domains work interconnectedly to create the Industry 4.0 ecosystem. All these domains are bridged by data both northbound and southbound and can be visualized as an analogous representation of the IDA-95 structure:

- Sensing, data prefiltering, and communications are part of the **Control** domain in the northbound direction. Southbound, the control module writes data and control signals to effect changes toward optimization.

- The **Operations** domain represents the collection of functions responsible for provisioning, managing, and monitoring the systems in the **Control** domain. The **Operations** domain sits one level above the **Control** domain, as it helps control and optimizes multiple plants in the enterprise, unlocking new business models.

- The **Information** domain is responsible for managing and processing data. The data collection and analysis functions built within this domain complement those in the **Control** domain. While the functions help in actuating physical devices in the **Control** domain, the **Information** domain functions help in decision-making and improve the system models horizontally.

- The **Application** domain holds business-specific application logic, rules, and models for optimization at a global level. It also contains APIs and UIs for data exchange between other layers and applications.

- The **Business** domain integrates business functions, enterprise resource planning, product life cycle management, human resource management, planning, and work scheduling systems, budget and financial systems, customer relationship management, and service life cycle management.

Apart from the major domains, we have **Crosscutting Functions** or enablers that help make solutions work. For example, **Connectivity** plays an important role in bridging multiple system functions. Other examples include **Data management**, **Analytics**, and **Intelligent control**. At the top plane of *Figure 9.3*, we have the **System Characteristics** or the architecture principles as we have explored in the initial section. These fundamental characteristics intertwine the enabling functions to build a trustworthy system.

Now that we have explored IIRA and its viewpoints, it's time to take a deeper look at RAMI 4.0. This provides an alternate view on the same topic, allowing the user to visualize the manufacturing of a product across all three dimensions, the life cycle, the hierarchy, and the process flow.

RAMI 4.0

The RAMI 4.0 is a service-oriented architecture model that combines manufacturing elements, business processes, and IT components in a layer and life cycle model. RAMI 4.0 is an abstract model for systematically mapping complex relationships and functionalities of an Industry 4.0 paradigm in a three-dimensional cube as shown in *Figure 9.4*. The cube has hierarchy levels, product life cycles, and value streams on the horizontal and vertical data integration layers. The idea of integrated manufacturing is that across the production chain, all the entities involved will have access to all the data necessary for the specific needs of each sector.

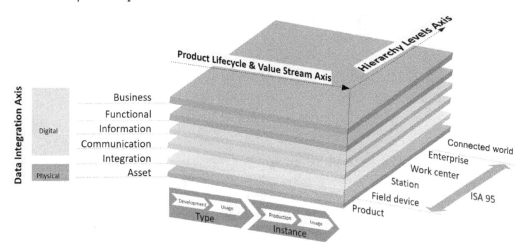

Figure 9.4 – RAMI 4.0 cube

Let's now break down the complex-looking three-dimensional cube and analyze each plane. Each plane represents a complete workflow in itself. It is interesting when they all come together to form a 3D cube. As we go through each of the three planes, readers can pause and visualize the corresponding components on the other two planes for each layer. An example could be a field device (e.g., the PLC) used in a manufacturing line with its asset production cycle from OEM, and so on.

Axis of hierarchy levels

IEC 62264 and IEC 61512 are technical standards defining this axis, consisting of seven components from the smart product to the connected world, as depicted in *Figure 9.5*.

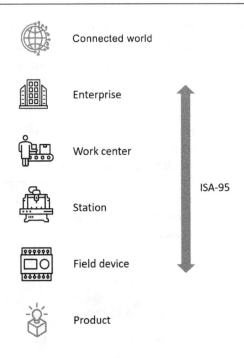

Figure 9.5 – Product development access

The axis can be seen as the north-south extension of the ISA-95 model in a factory, with the smart products at the bottom and the connected world up top. The factory is not an overlay of layers but a network of interactions between intelligent devices and the connected world. Since we have already analyzed the five layers of the ISA-95 in depth (the factory) in *Chapter 5, OT and Industrial Control Systems*, we will focus on the other two: (smart) products and the connected world:

- **Smart product**: This denotes the entity that is getting manufactured. This device has built-in intelligence, which implies that it can specify itself and locate itself uniquely. It can also include intelligence to monitor health, operations, performance, and surroundings, sometimes taking control actions on its own or reporting back to the OEM data center for further action. Being accessible and traceable can be leveraged for use cases, such as software over-the-air updates, dynamic on-field calibration, and so on.

- **Connected world**: Aggregates manufacturing facilities and their integration with external supporting enterprises, third-party suppliers, customers, partners, sales units, distributors, and system integrators. For example, the supply chain can be accessed to deliver raw materials requested by the production line on a need basis (demand-supply mapping). Another example would be to allow remote access for a component supplier to remotely debug production issues at short notice.

Axis of product life cycle

This axis, as pictured in *Figure 9.6,* tracks the development of a product from idea to end of life. This has two critical divisions, **Type** and **Instance**:

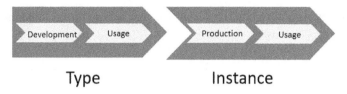

Figure 9.6 – Product life cycle

Type involves the product development phase, consisting of prototyping, computer simulation, and so on that helps build the product. The **Usage** module within **Type** specifies the standard maintenance cycles, operating procedures, instructions for operation, and so on. Once the product is designed, it goes in for mass production. Each **instance** of the product is then manufactured and labeled uniquely with an identity number (serial number). All product-specific information is bracketed into the **Production** component of the instance. The **Usage** component tracks the usage, wear and tear, maintenance information, recycling, and scrapping information.

Axis of layered data architecture

The third or vertical axis focuses on integrating the asset and business layers through data. This is represented in *Figure 9.7.*

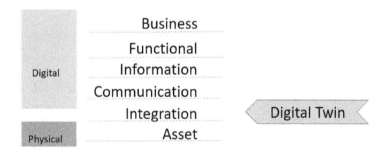

Figure 9.7 – Layered data architecture

In the bottom-most layer, a given **Asset** is represented in its **Physical** form. In the **Integration** layer, it is transformed into its counterpart, **Digital Twin**. The **Communication** layer connects various sources to the data lake, whereas the **Information** layer takes care of processing and obtaining valuable information. This information is mapped to the functions of the asset, where KPIs and interdependent metrics are calculated. Finally, the outcomes are presented to the **Business** layer in monetary correlation with production outputs.

Presenting architecture designs from two leading schools of thought provides the readers with a holistic understanding of how architecture frameworks are built for IIOT use cases. It is no longer about one use case for an isolated machine but a unified comprehensive design that interacts multi-laterally with horizontal and vertical systems in near real time to create value chains.

Let us now focus our efforts on two important concepts that help architects build modern, best-in-class designs for their applications:

- The AAS

- The DSB

Getting to know the AAS

In a nutshell, the AAS is a digital representation of a physical asset. It contains all the information and functionalities of the asset and acts as the segue between the physical asset and the connected digital world. There are provisions to describe and update parameters such as features, characteristics, properties, statuses, measurement data, calibration data, and capabilities.

The structure of the AAS is defined through a technology-agnostic meta-model and several technology-specific serialization mappings such as XML, JSON, and OPC UA. The meta-model can be composed of multiple submodels that update in real time in coordination with the state of the asset itself.

The AAS can also be considered the digital twin implementation for Industry 4.0. This implies a meta-model of every asset constantly updated using the IoT data from the asset. This meta-model can be used as an interface to interact and exchange information with other twins and systems. A simplified representation of the AAS is presented in *Figure 9.8.*

Figure 9.8 – The AAS

This topic is vast in terms of definition specifications and implementation details, so we will introduce the key concept of how an AAS works for brevity's sake. The AAS can be implemented for both intelligent and non-intelligent assets (through the IIoT) and is the standard to establish cross-company interoperability. It covers the entire life cycle of the asset and can form the digital basis for autonomous systems and AI.

The AAS meta-model for a motor asset is depicted in *Figure 9.9*.

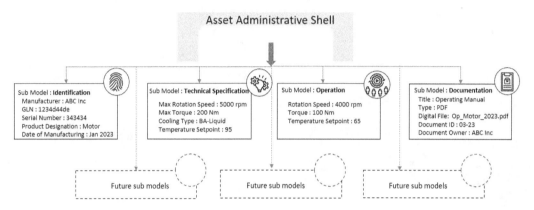

Figure 9.9 – AAS metamodel

The meta-model consists of several sub-models containing elements, functions, and values. In the case of the given asset, we have four sub-models that describe the asset: **Identification**, **Technical Specification**, **Operation**, and **Documentation**. The **Identification** sub-model is read-only and provides an immutable reference for the asset. The **Operation** submodel contains the current operating variables of the asset, presenting the real-time status. Further submodels can be added in the future. This model can then be technologically codified in XML or JSON format, packaged into an `aasx` package, and distributed to all the stakeholders in the value chain. The benefit of using this model is to standardize across enterprises and even across industries, which can lead to a product standard for sharing, as technical specifications as part of requirements for suppliers.

Getting to know the DSB

The DSB is a superset of a manufacturing service bus. The bus provides real-time data exchange between all parties involved in the planning production, distribution, service, insurance, and so on. This multi-lateral and collaborative data-sharing principle is fundamental to creating transparent value chains, as shown in *Figure 9.10*.

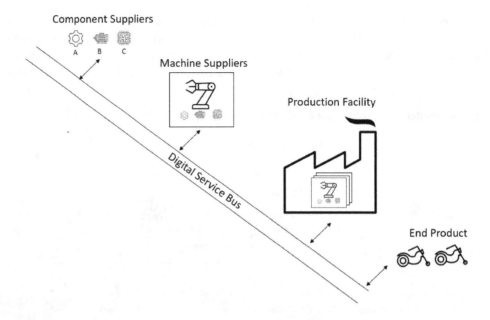

Figure 9.10 – DSB

Data availability in real time enables us to visualize supply chain fluctuations, respond to them, and minimize risks in supply chains. This central data highway also helps hone data-driven **Root Cause Analysis (RCA)** for any issues in the field. Actual operations and performance data, if available to the manufacturer, can help optimize product design and take corrective actions in case of future production.

DSB use case

Imagine a brake failure when a bike is driven at 150 kmp/h on a highway. Luckily, an accident is avoided, but the disgruntled driver tows his motorcycle to the workshop. When there is a DSB, all the stakeholders, the bike OEM, the component supplier, the system integrator, and the insurance company are immediately aware of the incident, with relevant information provided in a structured manner. Through data-driven RCA, active part traceability is faster, and decisions can be taken faster. Not only this but all active bikes on the road with similar system parameters can also be alerted to avoid similar situations. The design can be immediately improved for future production. Transparency, powered by data availability, creates trust across the entire value chain.

Practical considerations need to be incorporated when designing a solution architecture. While there is no quantifiable mechanism to define the quality of architecture, bear in mind that each solution, scenario, use case, and environment calls for a unique set of requirements around which architecture must evolve. *Evolve* is the keyword to be underlined, as this is a continuous process of system design and improvements. If the architecture checks the right boxes regarding the principles and patterns, rest assured we are moving in the right direction.

System considerations for architectural design

Now that we have a fair understanding of architectural principles and patterns, we would like to explore some key considerations that help drive architectural decisions. These are listed across the edge and cloud in *Figure 9.11*.

- Bandwidth/speed
- Data volume
- Network latency
- Privacy
- Security
- Autonomy
- Cost

- Availability
- Enterprise integration
- Global plant network
- Geo-political boundaries
- Compute power
- Storage
- Audit and governance

Figure 9.11 – System considerations

We have a clearer view of the edge and the cloud in moving decision-making's compute power and autonomy near the assets or to a central system. Bandwidth plays a very important role in system design.

As we know, the volume of data generated in the OT layer at the machines and sensors is very high and keeps compounding. Still, since most shop floor locations are remote areas with no access to high-speed internet, data transmission becomes a challenge. A strong edge with a higher storage capacity is recommended in these scenarios. We can then adopt a store-and-forward approach for the computed results or filter and forward to only send essential data (outliers, KPIs) to the cloud. In use cases where network latency is too high for decision-making, computing is offloaded to the edge.

Storage capacity with the mass collection of machine data from multiple sensors can quickly consume hard disk space when real-time data is needed for analysis. For instance, the continuous monitoring of a **Computerized Numerical Control** (**CNC**) machine at present requires about 1,500 measured values per minute, which corresponds to about 14 GB per year. This, when scaled to 300 machines across the shop floor, implies a scale of more than 4 TB in a year. When real-time monitoring, more than 70,000 measured values per minute would be required, which can very quickly fill the storage space. There needs to be a clear and intelligent data purge strategy or archival strategy to keep the edge running efficiently.

There are also cases where data is susceptible, and stakeholders do not like to risk sending data out of the plant network. In these cases, edge intelligence also takes precedence for the compute phase. The same applies to geo-political boundaries where governance restrictions exist on taking data from a particular location or country.

For large enterprises with hundreds of manufacturing facilities across the globe, integrated cloud architecture makes sense, as aggregated data can unlock multiple business value propositions. There have been instances where plants compete (in a healthy way) to optimize OEE to win the coveted best manufacturing award of the enterprise.

Besides all these operational factors, financial factors play an essential role in system design and long-term usage. A conscious decision based on quantifiable predictions needs to be made regarding upfront costs and running costs to arrive at an option viable for the long run.

> **Data is the new Sun**
>
> Data has become the single most crucial factor in any digital transformation project. Use cases, inferences, improvements, and optimizations are all based on data, but what types of data? Data can be classified in three ways in the context of IIoT – system, performance, and core data. *System data* contains information about the state, configuration, and possible alerts. *Performance data* is structured telemetry data about the system. This is used to keep a tab on the status and monitor the devices' working health. *Core data* is usually the key payload generated by the machines. These are the phy-gital (physical + digital) values emanating from sensor streams, energy meters, or cameras. The type and frequency of data must be considered for designing robust architecture frameworks.

Moving forward, we can consider analyzing a reference architecture focused on AWS. It congregates our learning so far in terms of design and applies it to a typical intelligent manufacturing use case with a compute-heavy edge and tightly knit cloud integration.

AWS reference architecture for IIoT

The AWS reference architecture for IIoT provides a good starting point to understand the construct of a solution and then build our own based on the project's specific needs. To emphasize, there is no single one-size-fits-all architecture solution, even though the principles and guard rails are universal. The reference architecture consists of familiar building blocks. The given **source** that provides the raw data from physical processes is aggregated at the edge of the **Collect** block. It is then transmitted securely to the cloud. The first block, **Ingest**, consumes the incoming data. Transformations and algorithms are applied to this data in the **Transform** block and stored in databases. The inferences and results obtained because of the application of algorithms are presented to the user in the **Insights** block.

The AWS reference architecture for a typical IIoT use case is presented in *Figure 9.12*.

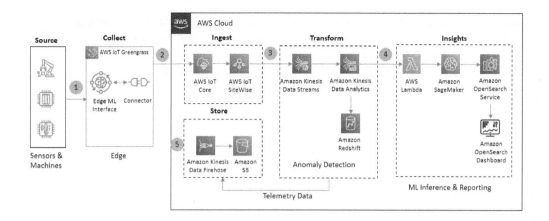

Figure 9.12 – AWS reference architecture for IIoT

Data from sensors, PLCs, and machines on the shop floor are interfaced with the edge computing unit through agreed protocols. The telemetry data is collected using AWS Greengrass running on the edge. Data is then ingested in the AWS cloud ecosystem using a managed cloud platform such as AWS IoT Core or AWS IoT SiteWise. IoT Core helps with the seamless connection of the edge to the cloud, providing an easy and secure path for data transmission. IoT SiteWise enables the collection, modeling, analysis, and visualization of IoT data at scale in a simple manner.

Kinesis Data Streams and Data Analytics process the stream data, transforming and analyzing using the Apache Flink and Beam frameworks. The Redshift service stores the resultant data for business intelligence reporting downstream. Data is also stored in an S3 data lake for further batch analysis and inferencing. For use cases involving ML, there is an EdgeML service at the edge and SageMaker in the cloud. The full-fledged ML service can train, deploy, and run the model to arrive at inferences and pass them on to consuming services.

Let's focus on the current point of interest in the industry – the applicability and practicality of AI to assist human intelligence on the shop floor. AI is here to stay as it stands today. Still, a definite, study-based approach weighing up all parameters of feasibility, viability, and desirability must be conducted before full-fledged investments are made.

Leveraging architecture for AI and ML

AI is a key enabler for value creation in smart manufacturing. AI is positioned to help transform rigid, static, and pre-defined processes into data-driven, demand-flexible, and dynamically adaptable processes. AI can also help create and transform products, services, business models, and processes by bringing high adaptability and computational ability in real time.

The ISO/IEC 2382 defines AI as a branch of computational science dedicated to the development of data processing systems that perform functions generally associated with human intelligence, such as logical reasoning, learning, and self-improvement. While it is not intended to copy human behavior, it is meant to increase the efficiency and effectiveness of industrial processes.

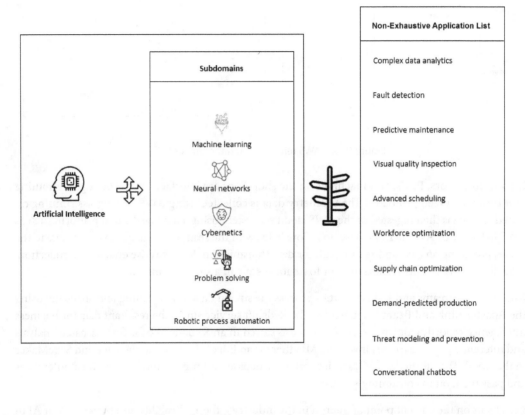

Figure 9.13 – AI ecosystem focused on IIoT

Figure 9.13 denotes the ecosystem for AI. While AI is a collective term, various sub-domains make up AI. These include but are not limited to ML, neural networks, cybernetics, AI problem-solving, and robotic process automation. Specific to the IIOT domain, many applications are not only bleeding-edge but have proven business potential in terms of faster ROI and VOI. The application lists featured are non-exhaustive and evolving every day. There are many possibilities, but we will focus on specific use cases with the most practical value.

In today's context, the practical use of AI in manufacturing is mainly twofold. The first is increasing production by optimizing machine availability, resource planning, and orchestrating processes around it. The second is to reduce downtime by predicting machine failure scenarios and preventing them through prescriptive maintenance.

ML capabilities are generally used to solve problems or derive inferences based on information extracted from an environment that is very complex or a multi-variable ecosystem where conventional optimization algorithms cannot accurately apply. We will dive deep into the applications of machine learning in smart manufacturing through practical AWS-based examples in *Chapter 12, Advanced Analytics and Machine Learning*.

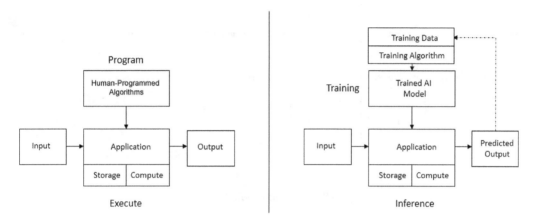

Figure 9.14 – Classic computing versus ML

Figure 9.14 shows a side-by-side comparison of classic computing versus computing with ML. In a classic computing scenario, the computations are static and codified by humans. This application is executed on nominal hardware that processes the inputs, calculates the output based on pre-defined formulae and logic, and expends the output. In the case of ML, an ML model is first trained using a specific algorithm and a training dataset. The trained model is deployed on compute-intensive, capable hardware. This model is run with real-time inputs, and output inferences are predicted. The predicted output is again used to fine-tune the model for accuracy.

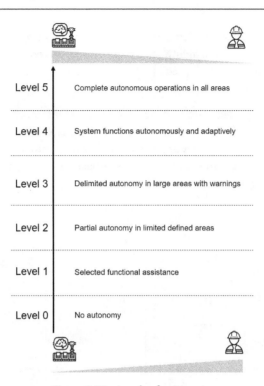

Figure 9.15 – Levels of autonomy

Figure 9.15 describes the various levels of autonomy between a human operator and an AI inference engine that controls a machine or production line. **Level 0** has a human at the helm and there is no control from AI. As we progress across the levels up to 5, the machine takes over and runs the entire operation while the human is a mere spectator. While the highest state is **Level 5** autonomy, this is not the most desirable solution in terms of today's evolution of technology and how humans trust computers. An assisted AI where the human still controls the decision-making and the AI suggests optimized actions in a timely manner is the way forward.

From compute-intensive edge resources, we have advanced into new technologies such as the FlatBuffers format access technology and edge GPU accelerators, **Tensor Processing Units** (**TPUs**), and **Intelligent Processing Units** (**IPUs**), making it more than lucrative to run inference engines at the edge. These topics define how complex computing can be distributed at the edge for decentralized decision-making, which is also a network-latency-independent response. A more detailed account is beyond the scope of this chapter, but readers are encouraged to explore this further.

Edge and cloud reference architecture for ML

Let us consider a sample reference architecture for an application leveraging ML, as shown in *Figure 9.16*.

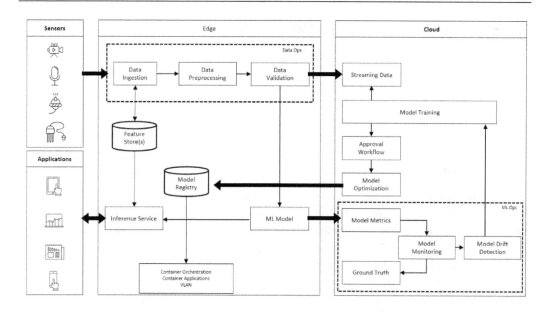

Figure 9.16 – Edge and cloud computes for ML applications

The use case is an integrated quality management system at the end of the production line. Due to the labor-intensive nature of the process, cameras are introduced that stream image frames of the finished product from multiple angles. The cameras are complemented with other aiding sensor data. This data is sent to the edge computing unit as part of the data ingestion process. The data is then preprocessed and validated. These three processes are together called edge DataOps.

Once the data is conditioned, it is sent as input to the ML model to perform the prediction operation. Simultaneously, the data is streamed to the cloud. The ML model at the edge leverages the inference service to publish the output data to the end user application. Various applications on the shop floor consume this data for logging, visualization, operator alerting, and so on, using multiple end devices, such as historians, dashboards, and mobile displays.

After the operation is completed, the ML model metrics are sent to the cloud. The model monitoring service compares the prediction with ground truth. If the cloud service identifies a model drift, model retraining is instantiated. This block of ML operations is termed **MLOps**. The ML model is then rebuilt using the IoT data that was received and optimized for the edge. The new ML model is then securely pushed to the edge using the ML data pipeline and deployed successfully. This process is repeated until the model is optimized for the given use case.

These processes might look simple (at least in theory!) to understand, and it may seem simple to appreciate the role of AI algorithms in the manufacturing domain, but systems are not perfect. Some imperfections again underline the importance of human-centric ML. Let us explore the most common issues with AI plaguing the industry-wide acceptance and rollout.

Common issues with AI for Industry 4.0

There are a few challenges within the large-scale adoption of AI-based use cases on the shop floor. Let's explore some of them and potential solutions for each.

As we know, training data is fundamental to an ML algorithm's success (achieving high recognition rates). Access to a large quantity of, as well as high-quality, training data, is problematic in many instances for any specific application. One possible workaround for this problem could be the use of transfer learning. Here, we begin with a pre-trained network from the same application class available open source, which is then retrained for the specific application with new datasets.

Another foundational issue with AI algorithms is the lack of *explainability*. Humans are wired to reason and thus associate a cause with a particular inference. Still, output from an AI engine may not be explanatory or logical to the human mind – in this case, the applicability and usability of a solution come under the radar. This is more of a philosophical dilemma where we need to decide upon the non-correlation between the results and the reasons. AI applications still do not achieve 100% coverage and cannot yet be relied upon as a replacement for a human. They should be viewed as an aid for making data-driven informed decisions where complex parameter calculations are involved. Experience and domain knowledge always hold the upper hand as we evolve into a collaborative man-machine workforce.

Thus, architecture decisions must be made, keeping in mind the limitations and requirements for a particular use case and its applicability rather than relying on general scenarios.

Summary

This was a chapter dedicated to architecture. Although we tried to condense many thoughts into a concise chapter, we explored various functional requirements and NFRs for architecture design. We also explored famous architectural patterns focusing on the manufacturing industry based on its fundamental principles. The five S principles of security, standardization, serviceability, scalability, and simplicity should be considered for any architecture; although there are other important NFRs, they tend to be a subset of these five.

We then delved into two widely used architecture frameworks, the IIRA and RAMI 4.0. Both provide detailed insight into how systems thinking for architecture must be considered with the utmost priority.

After this, we moved on to incorporating AI into manufacturing optimization. This changes how software and hardware are being perceived at the edge and the cloud due to how data is acquired and processed and inferences are done. In summary, the pinnacle moment of AI in manufacturing would be to understand operational data, factor in the context and environmental conditions, optimize processes, and assist people through empowered decision-making to augment production.

The end-to-end smart manufacturing use case is where we culminate all the learning presented so far. We gave the case of an enterprise digital transformation initiative and how the system can be designed considering both brownfield and greenfield deployments and all the dimensions according to the RAMI 4.0 model.

It is time to get our hands dirty again with some more code. The next chapter will build on the edge computing aspect that we discussed in this chapter's final section. Starting from an introduction to computing near the data source, it is a hands-on chapter focusing on services from AWS at the edge. Get ready to be on the "edge" of your seats!

10

Intelligent Systems at the Edge

It is an exciting time in industrial computing and IoT overall. The advances in cloud services and smaller and cheaper hardware have converged to provide immense opportunity to move further into the 4th and, possibly, the 5th industrial revolution.

In this chapter, we are going to cover the following topics:

- Edge computing and AWS Greengrass

- Setting up our scenario

- Communicating between Greengrass components

- Sending data to the cloud

Our focus for this chapter is on the edge – that is, within the factory network itself. While we will still leverage the cloud and send data for processing, most of our activity will be local to the equipment so that we can be co-located with our data and the machines and systems with which we interact.

Technical requirements

We will perform technical hands-on data gathering directly from a simulated industrial system such as a PLC. We will be taking the overall scenario a little deeper to allow you to understand how you can build on the examples to get started on projects. The code is more complex, but we attempt to explain it thoroughly. Hands-on deployment requires more advanced experience with Python, Modbus, AWS, IoT, and Linux administration. As always, we aim to make everything as understandable as possible. We will try to avoid too much complexity, but hands-on is required at this level. We encourage you to read through this chapter and then try it yourself. Mistakes will be made, and that's how we learn.

You can find the code samples mentioned in this chapter at `https://github.com/PacktPublishing/Industrial-IoT-for-Architects-and-Engineers/tree/main/chapter10`.

Edge computing and AWS Greengrass

Edge computing is not a new topic, but it can be confusing with all the marketing hype surrounding it today. The term *edge* is not unique in computing, and as early as the 1990s, teams were working on bringing content and data closer to end users to provide speed and efficiency. Edge caching of data or HTML pages on the web has been around for at least 20+ years.

We will flip the paradigm and focus on getting computing power closer to the data source rather than the consumer. Specifically, edge computing in our context is to move your computing power and processors as close to your machines and equipment as possible. We will leverage AWS Greengrass as our edge computer node in this chapter, but before we do a deep dive into this technology, let's review some of the ideas for edge computing. There are many advantages, especially in IoT. Consider the following benefits:

- **Pre-processing of data**: When capturing large data streams from a system, it can overwhelm the system when retrieving and processing the data if it's not designed for the expected volume. It may be the case that you don't need all of your data sent to the cloud, resulting in unwanted bandwidth and storage costs. Edge computing can help you pre-process that data and avoid unnecessary bandwidth and storage costs.

- **Faster processing time**: We use the term real-time data, meaning that we act upon or visualize data immediately when it becomes available. We often couch the concept using the phrase "near real time" to infer slight delays in the process. These delays are often caused by network latency or bandwidth issues between the plant and the cloud. Moving the processing of data directly into the plant (the edge) can help to avoid any delays for critical processing that needs to take place.

- **Avoid unplanned downtime**: With edge computing systems in place, we can decouple some of our dependency on the cloud or remote systems. Suppose something happens to our central processing systems or external network; edge systems can continue to run, store data locally, and perform some limited analysis, allowing plants to continue operating efficiently.

- **React intelligently to local events**: The goal should focus on connectivity and collecting relevant data, which has been the focus of this book. But once data is moving smoothly, you can jump to intelligent data analysis, starting with simple rule-based alerting and advancing to condition-based monitoring and machine learning.

There are also some potential disadvantages that you will see. The added complexity is one. Processing data at the edge can be compelling, but at the cost of additional hardware and software and building the pipelines and connections to the data. There may also be latency concerns in eventually getting data to the cloud. Although this depends on whether you are analyzing data within your edge device or only doing some initial processing for later analysis.

We are only scratching the surface when listing the advantages and potential disadvantages of edge computing, but this short list is essential for initial understanding. You cannot jump into edge computing without some basic knowledge; instead, wading in more carefully with a complete understanding of your goals is a better approach.

Technology options for your edge approach

Edge computing technology can take many forms – too many for full consideration, with many vendors coming up with a new strategy or approach almost every week. Options include using off-the-shelf solutions that work pretty well in some situations or more complex solutions using a framework or set of products.

Off-the-self solutions work well primarily when you are collecting data from similar systems: for example, if you are collecting data from well pumps or cold storage environments. In these examples, you have a similar technology set and are gathering the same set of measurements and data from each instance. Off-the-shelf technology often includes a set of adaptors and some programming or scripting language where you can set up connectivity and then process or forward the data, usually to a prebuild cloud environment with the ability to create dashboards. Take caution with this option since the dashboards may look pretty initially but may have less flexibility as you develop additional observational needs.

The advantage of this approach, however, is keeping things simple. In cases where you have many similar locations, you can build your integration and deploy it to hundreds or thousands of sites with minimal fuss. Solutions in this category can be hardware or software-based, depending on your use case or environment. The disadvantage can be when you have a very disparate environment or multiple environments, where a single solution doesn't always have all the connectors or functions you may need.

One alternative is to use a more open software framework, such as AWS Greengrass. Greengrass allows you to custom-build what you need and even go beyond what you might expect possible at the edge; this is the approach we will be exploring in this chapter.

Greengrass to the rescue

AWS IoT Greengrass provides an alternate approach for a comprehensive, supported edge connectivity and processing solution. Greengrass combines a runtime solution that can be installed and run on Windows or Linux hardware, along with cloud services that can communicate with and monitor the edge systems. Installing the Greengrass runtime or client on edge hardware is defined as an AWS IoT Greengrass Core device. This core connects to local devices and runs local processes called components on the local edge hardware.

Figure 10.1 shows how this looks in simple terms. This view represents the example we will build later in this chapter, with the main components outlined that we need to consider. We will expand on this view later in *Figure 10.6* to show in more detail how all the pieces will work together in our scenario; however, it is enough to understand some of the basics for now. We have seen AWS IoT Core already in previous chapters, so we know the essential points about receiving data from a device and processing that data. We will use the Greengrass section in AWS IoT Core, which allows us to deploy Greengrass edge devices and ensure they are set up and communicating with the cloud effectively:

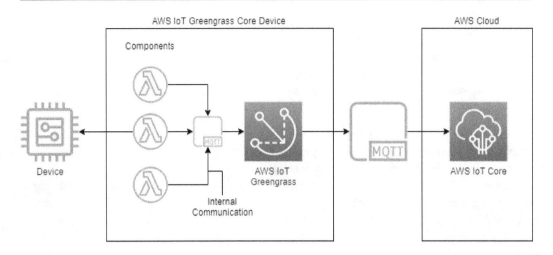

Figure 10.1 – AWS IoT Greengrass

The AWS IoT Greengrass runtime is a complicated piece of software. It has to be to ensure everything is in the right place. Uptime, monitoring, deployment, and security all have to be running at their best to provide a set of services in which you can be confident. In addition, the flexibility required to run all the different types of processes adds additional complexity, like a Swiss army knife at the edge. Much of this complexity is abstracted from us, but I mention it because becoming a Greengrass expert will take some time, and understanding the details will be beyond the reach of this chapter.

We will set up Greengrass and deploy several components to the core device for communicating with our factory machinery. You can consider these components as functions you deploy to Greengrass core devices that provide some processing. These components can invoke or provide communication to external devices, such as a Modbus TCP device, or invoke communication and processing for additional components if necessary.

Installing Greengrass

Installing Greengrass on a core device is not difficult but requires some attention. There are several different ways to install Greengrass core, depending on whether you are installing the software locally on the hardware, or running Greengrass within a Docker container, perhaps within a Kubernetes cluster running within the factory. Our example will involve installing it directly on a small computer, sometimes called an edge device. There are other options for learning and development, such as using an EC2 or Cloud9 instance. If you need a local machine, such as in our case for testing, then a Raspberry Pi or a small **Next Unit of Computing** (**NUC**) would be more than sufficient. In our use case, we want the device to be co-located or at least accessible within the same network as the equipment with which we will be interfacing.

Before we get too deep, it will be helpful to know where to start looking for more information as you go. AWS Greengrass has a reasonably comprehensive developers guide, which you can check out at `https://docs.aws.amazon.com/greengrass/v2/developerguide/what-is-iot-greengrass.html`. We will try and explain a few gotchas to be aware of in this chapter if you are a newbie.

We will start with the most straightforward installation: using the wizard and automated provisioning of the core. After reviewing the steps in this process and the output from the installation and setup on the device, you will better understand some of the underlying requirements for running Greengrass. Greengrass devices are located in the **Manage** section of **AWS IoT service**. Since we want to install a core device, start with that page and choose the option to **Set up one core device**.

Figure 10.2 shows the setup page for a single Greengrass core device. First, we need to replace the default name with something that makes sense to us. In our case, we will change the name of the device to Greengrass-103, which corresponds to the last octet of the IP address of our Greengrass device. There is no right or wrong here; use something that can allow you to track and manage devices in a meaningful way:

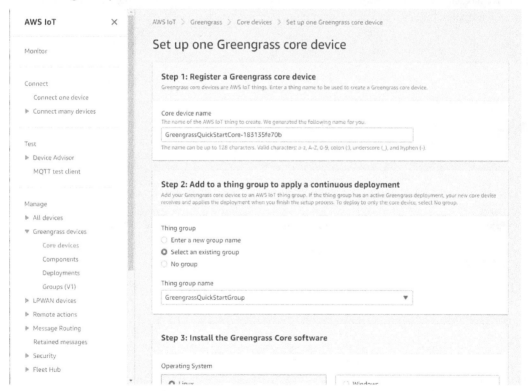

Figure 10.2 – Registering the device

You also need to set up a new **Thing group name** if you don't already have one. Thing groups allow you to group your core devices for management and deployment. For now, select **Enter a new group name** and enter a phrase such as `Greengrass_Edge_Connectivity` for the group name.

Now, we can start to set up the device to receive the Greengrass client software. *Figure 10.3* shows the pre-install steps. We should clarify that we are using Linux in our example – in fact, for all the examples in this book. Windows installs are also very popular but will not be discussed here.

Greengrass is a Java application that requires Java to be installed on the machine. Java may already be available on your OS, and you can check using the following command:

```
$ java -version
openjdk version "11.0.15" 2022-04-19
OpenJDK Runtime Environment (build 11.0.15+10-post-Debian-
1deb11u1)
OpenJDK 64-Bit Server VM (build 11.0.15+10-post-Debian-
1deb11u1, mixed mode)
```

If Java is unavailable, the setup guide has a link shown in *Figure 10.3* that provides the correct installation command, depending on your operating system:

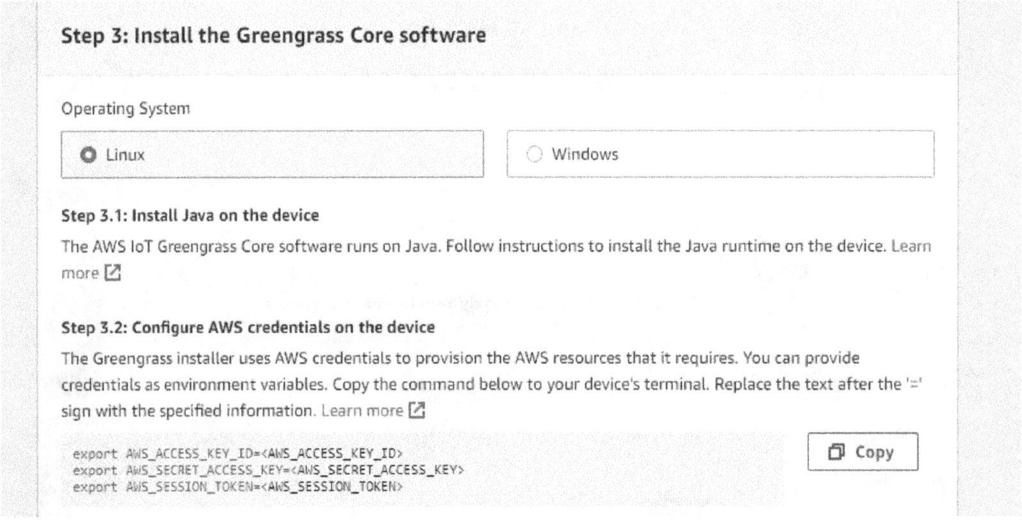

Figure 10.3 – Preinstallation setup

Step 3.2 in the installation process provides the commands for connectivity to the IoT Core service in the AWS cloud. It would be best if you defined the credentials to access your AWS account. This should look familiar to those who have used the AWS CLI previously. If not, the Greengrass installation process must add the proper credentials to the OS for access, as described on the page. The export command in Linux defines environment variables within the OS that other processes can use.

AWS credentials for installation

The credentials approach makes it easy to install AWS IoT Greengrass Core software. It allows the installer to connect to the AWS cloud and set up requirements for Greengrass to run. It is generally used for test environments to enable you to develop and test Greengrass configurations and components. Since it is assumed the machine is within your local sphere of control, you can use your own AWS credentials on this device.

There are additional ways to set up Greengrass without using your AWS credentials. For example, you can manually set up the required resources, IAM roles, and permissions within the cloud for Greengrass to connect without requiring your access credentials.

Finally, we can run the commands on the OS to install and configure the Greengrass client. Two commands are pre-defined for you in *Figure 10.4*. The first one is used to download the Greengrass client software. The second will run the installation process with all the necessary parameters required by the IoT Core installer:

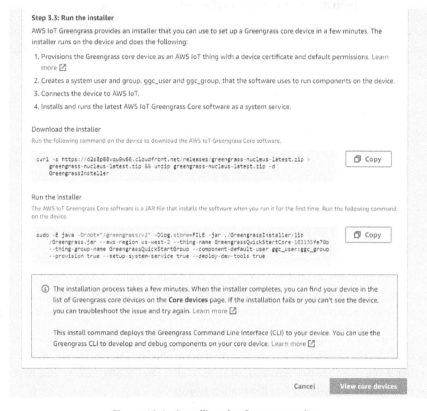

Figure 10.4 – Installing the Greengrass client

The hardware we are using is a simple Linux machine with hostname of gg103, and we are running all the commands as root in this example. A Raspberry Pi works perfectly for this to just get started learning at home. Automated provisioning is the simplest way to get things up and running, but we acknowledge that running things as root is not a strong security practice. Greengrass requires this because it makes low-level changes on the OS and sets specific Greengrass users on the system. Once you understand the environment better, you can impose stricter security measures on the runtime modules. Kicking off the installation with the provided command gets things going:

```
root@gg103:/greengrass# sudo -E java -Droot="/greengrass/
v2" -Dlog.store=FILE -jar ./GreengrassCore/lib/Greengrass.jar
--aws-region us-west-2 --thing-name Greengrass-103 --thing-
group-name Greengrass_Edge_Connectivity --component-default-
user ggc_user:ggc_group --provision true --setup-system-service
true --deploy-dev-tools true
```

We have included the AWS region and user and group parameters in the installation process. This is a default setup command that is provided by the Greengrass setup page and can be customized to fit your needs. However, it's probably not necessary to change things on the first try.

Once the process begins, we can view the output from the installation, as shown in the following result. Following along with the progress will provide a better understanding of what is happening during the installation process:

```
Provisioning AWS IoT resources for the device with IoT Thing
Name: [Greengrass-103]...
Found IoT policy "GreengrassV2IoTThingPolicy", reusing it
Creating keys and certificate...
Attaching policy to certificate...
Creating IoT Thing "Greengrass-103"...
Attaching certificate to IoT thing...
Successfully provisioned AWS IoT resources for the device with
IoT Thing Name: [Greengrass-103]!
```

The first part of the process sets up the IoT thing for our core device and configures the certificates and keys to ensure secure communication. Note that in this example, the installer found an existing IoT policy from a previous run, but when initially running, it will generate a policy with the default name.

For the sake of brevity, we will cut out some of the installation messaging here; however, we encourage you to review the process to understand everything that occurs better. Understanding the initial setup will help you later if you decide to perform manual steps or alter your configuration. This next section continues the output from the Greengrass installation process:

```
...
Created device configuration
Successfully configured Nucleus with provisioned resource
```

```
details!
Thing group exists, it could have existing deployment and
devices, hence NOT creating deployment for Greengrass first
party dev tools, please manually create a deployment if you
wish to
...
Successfully set up Nucleus as a system service
root@gg103:/greengrass#
```

We made it look easy here (I hope). But getting everything set up just right may take a few tries. In addition, you should consider a naming strategy for your Greengrass instances and set of use cases. This is important as you scale. It is not as important for one or two edge devices; however, when you scale to dozens or more, these little conventions can tremendously increase your environment's manageability. Consistency within AWS services, roles, and policies can make your life much easier. Write your standards and conventions down on paper and pin them to the wall. Consider this a best practice to ensure things are consistent as you grow:

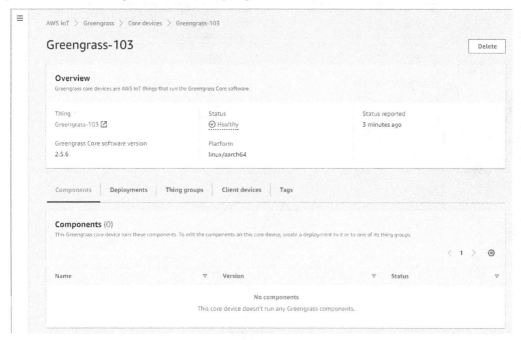

Figure 10.5 – IoT thing configured and connected

Figure 10.5 shows your Greengrass client connected to the cloud and running with a healthy status. You can review your setup and start preparing to deploy components to your Greengrass instance at this point. Next, we should examine the scenario we are working toward and understand how to use Greengrass to gather data from local devices.

Setting up our scenario

Our sample solution architecture is simple, and in line with most of the examples and use cases explored in this book. *Figure 10.6* provides an overview of the layered architecture we have discussed in detail. The core of the activity is in layer 2 of our overall diagram, where we focus on edge computing and applications:

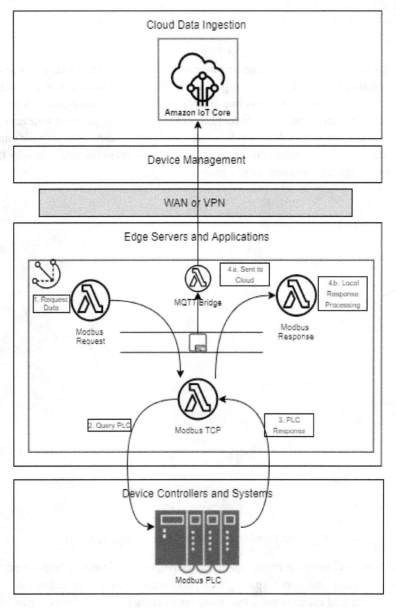

Figure 10.6 – Greengrass solution architecture

We have numbered the data flow and components within the environment to illustrate how functions are triggered, and how data flows within the activity sequence. We can describe the activity as follows:

1. **Request Data**: This component, called the *Modbus Request*, is a custom Python function that runs in a continuous loop. This process wakes up and triggers a message every few minutes, telling the Modbus component to query the PLC and retrieve some data.

2. **Query the PLC**: The *Modbus TCP* component performs both steps 2 and 3. Based on the request from step 1, or the ModbusRequest component, the Modbus TCP component queries the PLC via Modbus TCP and waits for the results.

3. **PLC Response**: A response comes from the PLC via the *Modbus TCP* component and returns the results as a new message on the MQTT broker.

4. Once the response is in from the broker, we have a couple of options we can consider:

   A. **Send a response to IoT Core**: We use an out-of-the-box component to send messages from the Greengrass message bus to IoT Core. The feature can be configured to send all messages or just what we need. Once the data is in the cloud, we can use our standard IoT approach for processing and analysis.

   B. **Process response locally**: *Optionally* we can build a custom component, *Modbus Response*, and deploy it locally to listen for answers and process them within the edge device.

These steps are logically grouped into phases of the process. Steps 1 and 2 retrieve or **pull** the data. Step 3 could be considered a **transform** function, and step 4 is a data **push** to move data to the next part of the ecosystem.

These steps are processed by custom-developed or configured components and then deployed to Greengrass to run locally. Keeping things real but straightforward, we want to provide examples that allow you to understand how Greengrass can retrieve and process data that resides outside the core, somewhere on the network. Each of the steps defined in *Figure 10.6* is processed by separate components that we build or configure and then deploy to our edge device. The components work together to achieve the flow of data that is outlined. In our example, we will deploy components that focus on each step in the process and align them to work together to retrieve the data.

Modbus data

In *Chapter 7, PLC Data Acquisition and Analysis*, we reviewed how we might set up a PLC to work with machines in the real world. Part of that interaction understands how things can communicate with devices and provide monitoring and control. The Modbus protocol is a data communications approach that is a long-time standard in industrial controls. To provide a realistic example, we will use a Modbus simulator to replace the PLC for testing and development. It would work the same if you were to use a real PLC; however, we wanted to look at an example that most of our readers could replicate easily on their workbench:

Figure 10.7 – pyModSlave simulator

Figure 10.7 shows an example of a simulator called pyModSlave, `https://pypi.org/project/pyModSlave/`, which is a free Python-based implementation of a Modbus slave application for simulation. This simulator is a Windows runtime that is quite simple to install and configure. We tried several free simulators, all of which worked the same, except that the registers' addressing seemed different for each. This does not seem uncommon as it is often different for different types of PLCs.

Modbus has been around since the late 70s and, as such, is well-defined for interaction between devices. Modbus has objects that are defined with a specific address space or register. Several Modbus *objects* or register types are pre-defined within the specification. A *coil*, for example, is a binary field that holds an on/off value. It is a read/write address of 1 bit. *Discrete input registers* are similar, except that they are read-only addresses. For more significant data values, the specification defines *input registers*, which are read-only, and *holding registers*, which are also read/write addresses.

The addresses of these registers are defined via the protocol. Still, they may differ, depending on your device, so some trial and error may be necessary to determine which address to request. In any case, the register addresses will be specific to your environment and configuration. In almost all instances, IT and OT should work together to determine the correct address to request, and what the value corresponds to in the real world (metadata).

Greengrass permissions

A quick note about Greengrass permissions. When Greengrass needs to do something within your AWS account, such as write logs to CloudWatch, or look up a value in Secrets Manager, it will require permission. As you know from previous chapters or your own experience, AWS follows a deny-first approach to security:

Figure 10.8 – Greengrass service role

A Greengrass service role is where you can add the policies to allow your core devices access to your AWS account services. Another good example is publishing messages to IoT Core from your edge devices. You can determine the service account by going to **AWS IoT** and clicking on **Settings**. Now, let's start deploying some components and making Greengrass do some work.

Communicating between Greengrass components

This section aims to build and deploy a few components, pull data from our Modbus simulator, and send it to IoT Core in the cloud. This section is going to be a little like drinking from a firehose. It is just a lot of information poured into one chapter. Because of this, we will gloss over information that can easily be gathered from online resources and focus on the solution's core. Less technical readers will gain a lot from this as it bypasses some of the minutiae of deploying components. More technical readers will need to spend additional time researching some of the steps we outline, but not in much detail.

The basic steps we will take are as follows:

1. **Modbus Request component**: Build and deploy a component requesting data from the Modbus slave.
2. **Modbus TCP component**: Deploy a function to query the simulator and retrieve the data.

3. **Modbus Response component**: Look at two options to retrieve the simulator data and begin processing.

We mentioned earlier that components in Greengrass are functions you deploy that perform actions. Components can communicate in several ways, but one of the primary approaches is to use **Inter-Process Communication (IPC)** within your component. The IPC library uses primarily an MQTT protocol that allows for the ability to subscribe or listen to a topic or optionally provides a request/response approach.

Components are broken down into **artifacts**, the runnable code, and **recipes**, which contains the deployment information. Together, those are deployed into Greengrass to allow a component to be installed and run as desired. Let's look at the first component and its respective parts.

Modbus Request component

The Modbus Request component will start the chain of events we are creating. It is a custom component that requests specific data from the Modbus TCP component. The Modbus TCP component then receives the request and goes about its business to query the Modbus simulator.

As mentioned previously, components are comprised of two elements – an artifact (in this case, Python code) and a recipe that tells Greengrass how to manage the component. The recipe is a JSON or YAML file, as you will see in a moment.

The code for this artifact is relatively simple, but there is a bit of setup in sending a message via ICP. Let's review the Python code for this component.

Modbusrequest.py (the artifact)

The code example for `modbusrequest.py` is Python code. Let's review the example:

```python
import sys
import datetime
import time
import logging
import awsiot.greengrasscoreipc
from awsiot.greengrasscoreipc.model import (
    PublishToTopicRequest,
    PublishMessage,
    BinaryMessage
)
```

All of the necessary modules are included in the preceding code snippet. Specific here are `greengrasscoreipc` and `greengrasscoreipc.model`, which allow us to access and send messages via the IPC, respectively. We import all the necessary libraries for sending an IPC message.

Next, we must set up a logger and define our wait time. The wait time is how long we wait for a response when sending our message via IPC:

```
# Setup logging to stdout
logger = logging.getLogger(__name__)
logging.basicConfig(stream=sys.stdout, level=logging.DEBUG)
FUTURE_WAIT_TIME = 10
```

Sleep time is how long we wait between sending messages. Essentially, the main program loop sleeps for this amount of time and then starts over and sends a message. We define this as 60 seconds to get one response for data every minute. This provides a consistent flow of data without overloading us when testing:

```
SLEEP_TIME = 60
```

Setting up our logging allows us to track what is happening within the edge device. We want to follow our activity locally at the edge to see what is happening, especially during the development and debugging phase.

> **Caution with excessive logging**
>
> Although we won't go into too much detail here, you should be cautious when sending logging data to CloudWatch. With edge devices, especially with many core devices, logging can become quite voluminous. Greengrass logging is a little tricky and doesn't seem to follow the same rules that developers are used to. This can quickly expand your cost expectations. We have seen this happen, and it can be pretty devastating. Just heed the warning: add alerts to your AWS account and be careful.

Next, we will instantiate our IPC client so that it can communicate with the service:

```
ipc_client = awsiot.greengrasscoreipc.connect()
topic = "modbus/request/conveyer"
#message = '{"request": { "operation":
"ReadCoilsRequest","device": 1,"address": 1,"count": 1 }, "id":
"1" }' # used for modbus-rtu component
message = '{ "id": "TestRequest", "function": "ReadCoils",
"address": 00001, "quantity": 10 }'
#message = '{ "id": "TestRequest", "function":
"ReadHoldingRegisters", "address": 400001, "quantity": 3 }'
```

```
#message = '{ "id": "ReadVoltage", "function":
"ReadInputRegisters", "address": 11, "quantity": 3 }'

logger.debug("topic: " + topic)
logger.debug("message: " + message)
```

Note that we have four different messages, with three commented out. The name of each function is defined by the ModbusTCP component, which can be found at https://github.com/awslabs/ aws-greengrass-labs-modbus-tcp-protocol-adapter. This is an awslabs component that is freely available for your use. It is well documented on the GitHub page where it is available. This next section is pretty standard for sending messages with IPC. It creates a set of objects that encode the topic and the message you are asking to send. We will send this same message every 60 seconds, so we generate the objects once and use them every time we want to send them:

```
request = PublishToTopicRequest()
request.topic = topic
publish_message = PublishMessage()
publish_message.binary_message = BinaryMessage()
publish_message.binary_message.message = bytes(message, "utf-
8")
request.publish_message = publish_message
```

Once we have our request object configured, we can forward the message. At this point, we go into an infinite loop. Again, since this is a simplified example, we send the request for data every 60 seconds, as defined in our SLEEP_TIME constant:

```
while True:
    operation = ipc_client.new_publish_to_topic()
    operation.activate(request)
    future = operation.get_response()
    future.result(FUTURE_WAIT_TIME)
    # Append the message to the log file.
    logger.info(message)
    print("going to sleep")
    time.sleep(SLEEP_TIME)
```

The loop is set up to go forever, sleeping for 60 seconds before it wakes up and publishes the message. The activate method sends the message and is followed by a future event with a declared wait time. This Future operation returns None if the process is successful; otherwise, it will raise an exception if the request fails. We can follow the logs locally on Greengrass Core to ensure that the message is sent and received correctly.

The example log files you can view in the following snippet are included for reference and can be tracked locally by using a `tail` command on the Linux command line:

```
2022-09-09T09:34:02.767Z [INFO]
com.environmentsense.modbus.modbusrequest: stdout.
DEBUG:awsiot.eventstreamrpc:<awsiot.greengrasscoreipc.client.
PublishToTopicOperation object at 0x7f8d8c33d0> sending request
APPLICATION_MESSAGE [Header(':content-type', 'application/
json',
<HeaderType.STRING: 7>), Header('service-model-type',
'aws.greengrass#PublishToTopicRequest', <HeaderType.STRING:
7>)] b'{"topic": "modbus/request/conveyer", "publishMessage":
{"binaryMessage": {"message": "eyAiaWQiOiAiAiUmVhZFZvbHRhZ2UiLCAi
ZnVuY3Rpb24iOiAiAiUmVhZElucHV0UmVnaXN0ZXJzIiwgImFkZHJlc3MiOiAxMSw
gInF1YW50aXR5IjogMyB9"}}}'.
```

```
2022-09-09T09:34:02.771Z [INFO]
com.environmentsense.modbus.modbusrequest: stdout.
DEBUG:awsiot.eventstreamrpc:<awsiot.greengrasscoreipc.client.
PublishToTopicOperation object at 0x7f8d8c33d0> received #1
APPLICATION_MESSAGE [Header(':content-type', 'application/
json', <HeaderType.STRING: 7>), Header('service-model-
type', 'aws.greengrass#PublishToTopicResponse', <HeaderType.
STRING: 7>), Header(':message-type', 0, <HeaderType.INT32:
4>), Header(':message-flags', 2, <HeaderType.INT32: 4>),
Header(':stream-id', 118, <HeaderType.INT32: 4>)] b'{}'.
```

```
2022-09-09T09:34:02.772Z [INFO]
com.environmentsense.modbus.modbusrequest: stdout. INFO:__
main__:{ "id": "TestRequest", "function": "ReadCoils",
"address": 0001, "quantity": 10 }.
```

Note that the message that's sent to the Modbus slave is a `ReadCoils` request. Modbus functions are reasonably well-documented and can be found via a quick Google search. This one is a Modbus function that essentially reads discrete inputs into the system. These are binary values, as shown in *Figure 10.7*. We expect this request to return the first 10 coil values in PLC. This should be a set of `true`/`false` values based on the current settings:

```
2022-09-09T09:34:02.773Z [INFO]
com.environmentsense.modbus.modbusrequest: stdout. going to
sleep.
```

This code will be new to anyone just starting with Greengrass, but it uses the standard IPC library for communication with components within the system. This component is developed as a single Python file, so it is one of the most straightforward functions to deploy to Greengrass.

Let's look at the second half of the component, which is a deployment descriptor called a **recipe** in the Greengrass vernacular.

Modbusrequest.json (the recipe)

A recipe is a file that defines how a component is deployed and run. It can be defined in JSON format or as a YAML file. The file for the Modbus request is set up as a JSON file and contains details about the program, dependencies, and the component's life cycle:

```
{
    "RecipeFormatVersion": "2020-01-25",
    "ComponentName": "com.environmentsense.modbus.
modbusrequest",
    "ComponentVersion": "1.0.0",
    "ComponentDescription": "modbus TCP Request component.",
    "ComponentPublisher": "EnvironmentSense",
    "ComponentConfiguration": {
      "DefaultConfiguration": {
        "Message": "request",
```

A recipe starts with standard information, such as its `FormatVersion`, name, description, and more about a file that defines how a component is deployed and run. Of crucial importance here is `ComponentVersion`. Greengrass uses semantic versioning that follows a *major.minor.patch* strategy; you will need to track the versioning of your components as you continue your development:

```
        "accessControl": {
          "aws.greengrass.ipc.pubsub": {
            "com.environmentsense.modbus.
ModbusRequest:pubsub:1": {
              "policyDescription": "Allows access to publish to
topic.",
              "operations": [
                "aws.greengrass#PublishToTopic"
              ],
              "resources": [
                "modbus/request/conveyer"
              ]
```

```
              }
            }
          }
        }
      },
```

We can define authorization policies with the `accessControl` configuration parameter. The `accessControl` section maps IPC services and authorizations available to the component. In this case, we allow our component permission to publish to the `modbus/request/conveyer` topic, where the Modbus TCP component listens for commands or requests:

```
    "Manifests": [
      {
        "Platform": {
          "os": "linux"
        },
        "Lifecycle": {
          "Install": {
            "script": "python3 -m pip install --user awsiotsdk"
          },
          "Run": {
            "script": "python3 -u {artifacts:path}/
modbusrequest.py '{configuration:/message}'"
          }
        }
      }
    ]
  }
```

The final section is the Manifest. The Manifest of our component has three areas:

- `Platform`: In this case, this is specific to the Linux platform.
- `Lifecycle`: This tells Greengrass to install the component, but it also depends on the **awsiotsdk**, which is needed to import the IPC libraries.
- `Run`: This tells Greengrass how to run the component. There is an optional configuration parameter, which is not used in our example.

This is the most significant code set for our example scenario, but we need to deploy our component to Greengrass before anything can happen.

Component deployments

Deploying components can be done in several ways – either through the AWS console or the command line. For initial development and testing, I favor using the command line and working locally on the Greengrass machine when possible. It is quicker and easier once you get the commands correct. Don't get too discouraged; even for long-time programmers and Linux administrators, this is a learning process. If you are like me, you keep different note pages on your desktop with a list of commands you commonly use.

> **Greengrass developers guide**
>
> Before we get too far, I would be remiss not to point you to the AWS IoT Greengrass developers guide. The guide for version 2 of Greengrass, as of this writing, is available at `https://docs.aws.amazon.com/greengrass/v2/developerguide`. This guide is a reasonably comprehensive set of information that can guide you through understanding some of the main features of Greengrass and deploying and running components.

Visual Studio Code (**VS Code**) is my weapon of choice for taming this beast. It has grown into a Swiss army knife of functionality, regardless of the language or platform. It allows SSH into your local edge device, editing your code, running terminal commands, and even providing port forwarding to access the local Greengrass debug console:

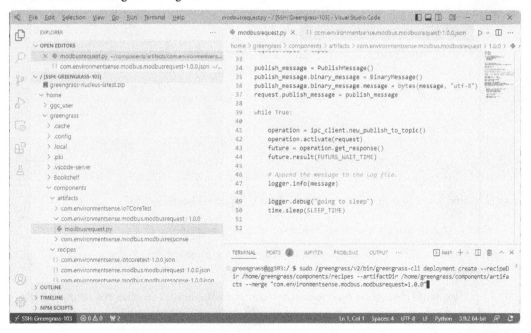

Figure 10.9 – Visual Studio Code SSH to Greengrass

Figure 10.9 is a screenshot of VS Code for this example. Even if you are experienced with VS Code, this is a handy extended setup. We have configured SSH within VS Code to connect to the device and can work on the code examples directly on the device for local deployment. In addition, we have added port forwarding for the local debug console. You can see two ports enabled on the port tab: 1441 and 1442. Using VS Code, you can work virtually locally on your Greengrass device without using the AWS Console for all your interactions, which can slow you down during development and testing.

There are two folders under the home directory. The first folder is for the artifacts and contains subdirectories with the artifact's version number. In our case, this is version 1.0.0. The second folder is for the recipes. This folder does not need subfolders but could be separated if you want to keep different recipe versions. You can deploy this component using the `greengrass-cli` `deployment` command:

```
$ sudo /greengrass/v2/bin/greengrass-cli deployment create
--recipeDir /home/greengrass/components/recipes --artifactDir
/home/greengrass/components/artifacts --merge "com.
environmentsense.modbus.modbusrequest=1.0.0"
```

If the deployment is successful, you can go to your Greengrass logs folder and view the log files as the component runs. The `tail` command is a popular command for viewing logs, and it will be available on any Linux system:

```
$ cd /greengrass/v2/logs/
$ tail -f com.environmentsense.modbus.ModbusRequest.log
```

Here, we simplified the deployment process while providing all the relevant information. Some trial and error will probably be necessary, but with some careful reading, you should be able to deploy your component with minimal changes. There are many references on the internet about deploying components, so do not worry if this seems like a lot to uncover:

```
$ sudo /greengrass/v2/bin/greengrass-cli deployment create
--remove "com.environmentsense.modbus.modbusrequest"
```

Now, we need something to pick up our message and act on it. This is the next component in our list, which is a prebuilt component that will talk to our Modbus simulator and retrieve the values we are requesting.

The next component – ModbusTCP

Greengrass components come in all shapes and sizes. Essentially, there are three major types of components.

1. **My components**: These are components you build yourself, just like the Modbus request component we created in the previous section. We developed and tested that component locally

on our Greengrass device; however, once we are comfortable with the code and function, we can move that more formally into the UI and deploy the component globally to all our edge devices.

2. **Public components**: Public components are provided by AWS and available within AWS IoT for use in your environment. No general code is available; however, you can change the recipe and configure these components in your environment.

3. **Community components**: These are components that are developed by the Greengrass community and are collected and stored by the AWS IoT Greengrass team. They are available on GitHub, and a link is provided on the component page to view and use this library.

The next component we will use is part of the open source community within AWS Labs. You can find this component here: `https://github.com/awslabs/aws-greengrass-labs-modbus-tcp-protocol-adapter`. This component allows you to read or write device data using the Modbus TCP protocol.

Why different components?

It's easy to get confused with different components with the word Modbus in the name. But each component has a different function, and only by working together can we achieve our goal.

The prebuilt Modbus TCP component allows us to query and retrieve data from our Modbus slave on the network. It has many pre-defined capabilities, enabling it to talk to the slave using the Modbus protocol. It cannot, however, work in a vacuum. It needs to be told what data to query and return via an IPC request. That is the job of the previous Modbus Request component, which wakes up every few seconds and sends a message to Modbus TCP requesting some set of data.

The main page for the Modbus TCP component offers a precompiled binary file for deployment into your environment. You should copy the referenced `.jar` file to an S3 folder for deployment to the edge. The recipe is provided as a YAML file, which needs to be customized with specific information required to access the Modbus system, described as follows:

```
{
  "Modbus": {
    "Endpoints": [
      {
        "Host": "192.168.178.34",
        "Port": "502",
        "Devices": [
          {
            "Name": "conveyer",
            "UnitId": 0
          }
```

```
      ]
    }
  ]
},
```

For this component, you must customize the recipe with the IP and port of your Modbus endpoint. The recipe allows you to provide that input for the IP address and port to access, as well as a name for the device. This device name is essential as it is used to send messages specifically to trigger this component:

```
"accessControl": {
  "aws.greengrass.ipc.pubsub": {
    "aws.greengrass.labs.ModbusTCP:pubsub:1": {
      "policyDescription": "Allows publish to all topics.",
      "operations": [
        "aws.greengrass#PublishToTopic"
      ],
      "resources": [
        "*"
      ]
```

We set up the access control for this component so that it's very open, allowing access to send and receive messages on any topic. The following section shows "*" access to resources or topics. This means we can listen to any topic within the IPC. We would not want to go into production with this setting. Once we have the component working how we like, we can lock this down to what is necessary:

```
    "aws.greengrass.labs.ModbusTCP:pubsub:2": {
      "policyDescription": "Allows subscribe to all topics.",
      "operations": [
        "aws.greengrass#SubscribeToTopic"
      ],
      "resources": [
        "*"
      ]
    }
  }
}
```

The full recipe is available in this book's GitHub repository, which includes the information for deploying the component to different systems. Note that our component listens on the `modbus/request/conveyer` topic, which our Modbus request component uses when it sends a response to the request message:

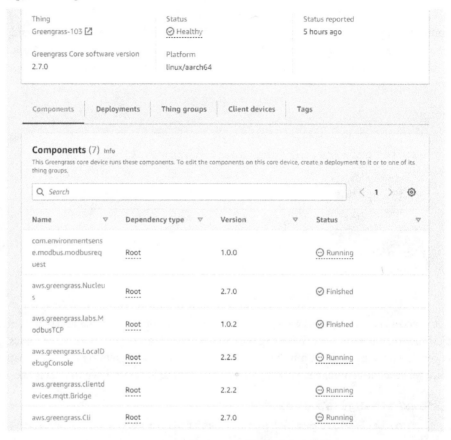

Figure 10.10 – Deployed components

Figure 10.10 shows the components that are running on Greengrass 103. Greengrass updates its status to the cloud even when components are deployed locally so that you can get a clear picture of what is deployed to any particular cored device. Notice that we have also deployed the `greengrass.Cli` feature, which allows us to use the `deploy` command shown earlier to deploy code locally for development.

In addition, the local debug console is also deployed, which has a web-based interface to view and edit deployed components. To use this debug console, you must enable port forwarding over SSH on your VS Code terminal to launch a browser window locally. Ports `1441` and `1442` must be forwarded. Then, you can open a local browser window to `https://localhost:1441/` to launch the console.

The console also requires a username (default: debug), and a password, which can be obtained using the get-debug-password command. This command returns the following output:

```
# sudo /greengrass/v2/bin/greengrass-cli get-debug-password
Sep 10, 2022 6:52:10 AM software.amazon.awssdk.eventstreamrpc.
EventStreamRPCConnection$1 onConnectionSetup
INFO: Socket connection /greengrass/v2/ipc.socket:8033 to
server result [AWS_ERROR_SUCCESS]
Sep 10, 2022 6:52:10 AM software.amazon.awssdk.eventstreamrpc.
EventStreamRPCConnection$1 onProtocolMessage
INFO: Connection established with event stream RPC server
Username: debug
Password: 2SJ4oF6ibB6xTIfvy3208AXD_RXEEHD0WLyfwQmzyUk
Password expires at: 2022-09-10T14:52:10.625413894-07:00
```

The console will tell you which components are running or whether they have failed. You can also update the configuration directly on the debug console and restart any failed components.

Whew! Take a breath here and realize what you have accomplished. We have deployed and connected to a Greengrass instance and configured multiple components to request data from a Modbus slave within the network. Our next step is to do something with that data – most importantly, sending it northward to the cloud for storage and analysis.

Sending data to the cloud

As with most things on the AWS platform, there are always several ways to design and build a solution; in this case, it depends on our ultimate goals:

- Process data locally on the edge? In this case, we need to build another custom component that listens to the correct topic for a response. The topic for this example would be modbus\
  response\conveyer, mapping directly to the name we provided in our ModbusTCP
  component configuration. Example code for this type of component is readily available in online examples and you can follow the same process we used for the request component.

- Forward data to the cloud? Suppose the only thing we need to do is forward the data to IoT Core. In that case, we can simplify our work somewhat by using a prebuilt component to deliver data to the cloud directly from local Greengrass IPC messages.

The reality is that it is probably a combination of these two goals we aim to achieve: process data at least somewhat locally and then forward that processed data to the cloud for additional storage and evaluation. The Greengrass MQTT bridge component relays messages between different areas, such as across clients or to AWS IoT Core.

In our case, we want to go ahead and relay messages from local pub/sub to IoT Core, and we can define which messages we want to send based on the topic by adding a mapping configuration to the component. The MQTT bridge component is an example of a public component that AWS IoT provides for your deployments. Since it is a public component, it requires additional customization to determine what data to send and where. For now, we'll make it simple – we will send all messages from the local system using pub/sub to the cloud:

```
{
    "reset": [],
    "merge": {
      "mqttTopicMapping": {
        "AllLocalMessages": {
          "topic": "#",
          "source": "Pubsub",
          "target": "IotCore"
        }
      }
    }
}
```

Most (all?) of the public components require some customization, either permissions or some parameter settings to enable the functionality you desire. Fortunately, it is not difficult, but it also doesn't seem to be well documented at the moment.

Figure 10.11 highlights the configuration button for a component. During deployment, you can select a component and then view the configuration for that component:

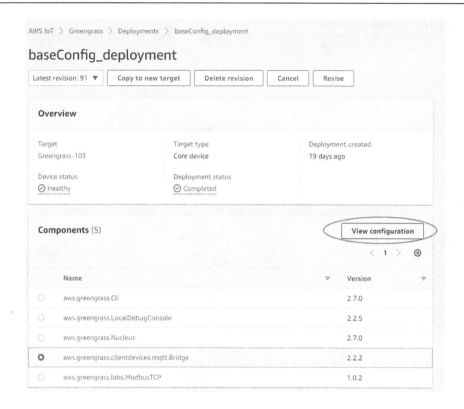

Figure 10.11 – Configuring a public component

This can be important because most public components will not work without some additional configuration for your environment. You can define the recipe directly and add the necessary information for custom or community components.

It can also be a way to update a configuration if you need it. For example, if permission is necessary to listen on a specific topic, using a wildcard parameter is not always the safest approach for large-scale industrial edge deployments. This raises the question of the zero trust or least privileges approach, which has become reasonably necessary today. Using the wildcard or open permissions is OK for development and testing, but locking down your access should be the next step before production deployment:

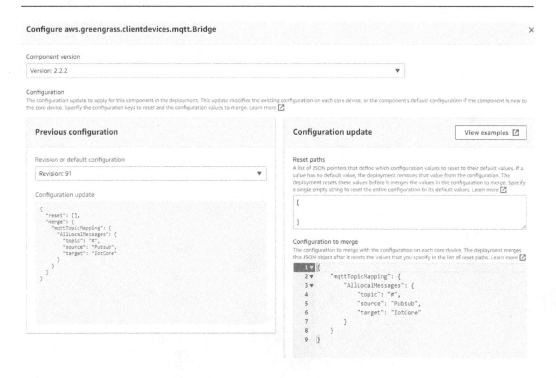

Figure 10.12 – Configuring the component during deployment

The preceding screenshot shows the configuration changes in place. These updates will be merged with the recipe to ensure the MQTT broker will relay the correct messages. In this case, all the messages will be forwarded so that we will receive both the request and the response in the scenario. Note that this component returns a value on the `modbus/response/conveyer` topic, which our ModbusTCP component was configured to initially.

Once the MQTT bridge component has been deployed, it should send messages to the cloud. You should check the permissions for the Greengrass service role we discussed earlier if you don't see data. Adding a policy to send data to IoT Core should open things up:

Figure 10.13 – AWS IoT MQTT client

Since we opened up the bridge component to listen to everything, we should see both the original request and the response shown in the preceding figure. This is a simple coil request, which is perfect for your first time. Each bit represents an on or off value for your production system. Depending on your use case, you can now perform some logic or machine learning to set values if something changes. The ModbusTCP adaptor we are using has several write commands, such as `WriteSingleCoil`, to support this effort.

Sparkplug protocol

We would be remiss not to mention the Sparkplug protocol and working group at Eclipse. The data we are sending to IoT Core is pretty straightforward MQTT data. The topic tells us where the data is from, and then embedded within the message JSON are the actual values and any available additional information. Sparkplug, managed by the Eclipse Foundation at `https://sparkplug.eclipse.org/`, adds an extra layer of information within the MQTT protocol. This includes more detailed namespace and state management.

AWS IoT Core does not currently natively handle the Sparkplug protocol, which encapsulates or compresses the data message. There are ways to handle this and AWS SiteWise is another alternative for managing this type of messaging. Unfortunately, we are not able to go into detail, but this is worth exploring on your own to understand the advantages this might bring.

Summary

We packed a ton into this chapter, providing an introduction and overview of edge computing and building an example to learn firsthand how everything works. More than likely, you are already thinking about your specific scenarios and itching to get started. Let's take a moment to consider what we have accomplished.

First, we deployed AWS Greengrass to an edge computer or device and then connected it to the AWS cloud to help manage the device. Then, we developed and deployed several components – one of each type of component: a custom-developed component, a public component, and an out-of-the-box Greengrass community component. The way each of those works is slightly different, but they work well together to accomplish our goals. Finally, we tied everything together to request some Modbus data. We queried a remote Modbus slave for the data, and then returned this data and forwarded it to the cloud.

We went beyond the basic HelloWorld example in this chapter; we wanted something that takes you into real-world territory. As such, we had to make some sacrifices in some detail but tried to provide enough pointers so that you could fill in the gaps. The takeaway was to give you enough information to try getting some actual data from your systems with just a little more effort beyond what is provided. We also showed an example of each type of component – custom, public, and community – and how they can interact to achieve a goal.

We have ignored some of the more complicated stuff, such as how to read other Modbus registers and decode the results. If that is your goal, there are multiple books and resources on interacting with the Modbus protocol. Later, in the next chapter, we will revisit our work and pull some more exciting data using Modbus. In the next chapter, we will look at remote monitoring and some high-speed monitoring situations to help you understand how to approach different scenarios by gathering data from more challenging environments and systems.

## 11

Remote Monitoring Challenges

Remote monitoring is fundamental to autonomous operations, and plays a significant role in business continuity and predicting failure scenarios and mitigating them. This chapter will focus on really remote situations and making decisions about bandwidth concerns, power consumption, and the volume of data being transmitted. Our focus will be on working with data acquisition in remote or isolated locations, as often faced by industries such as renewable energy and oil and gas. We will examine different options for data transfer, such as 5G, satellite, and long-range wireless options.

Through the aid of two case studies, we would like to drive home the system considerations, design decisions, compromises, technology selection, and processes the architect must undertake while working on remote monitoring applications. Our focus will also be to introduce to the readers the fourth dimension, along with desirability, viability, and feasibility on which architecture design must be based. Read through to discover more. We will be delving into the following topics in this chapter:

- The case for remote system monitoring
- Long-range data acquisition challenges
- Case study: remote monitoring of a solar farm
- Architecting for sustainability
- ESG case study: remote monitoring of a carbon capture and storage unit using IIoT

Remote monitoring has been one of the earliest success stories of digitalization – providing users from anywhere in the world with a mechanism to learn, understand, and control the behavior of a system.

Technical requirements

The objective of the chapter is to explore the concept of remote monitoring systems, current challenges, and design approaches to defining system architecture and overcoming these challenges. This chapter also tries to present an important dimension of sustainability when architecting solutions. While this is a conceptual chapter, you can follow along even without in-depth knowledge of the topics. But a generic understanding of the following areas will set you in good stead:

- Edge computing
- Wireless transmission technologies

Why do we need remote monitoring?

Remote system monitoring is the ability to monitor the behavior of a system located far from the user in a systematic and quantifiable manner. Remote monitoring has been enabled through the availability of operational data and in-situ sensor data in a centralized platform. Remote monitoring not only replaces manual intensive processes of recording system behavior and reacting to breakdowns after the incident has occurred to a more proactive and preventive approach in terms of service and maintenance. When the systems of interest are furthermore located at really remote locations (very far-off and inaccessible locations), this becomes a savior guaranteeing the continuation of operations.

The importance, relevance, and absolute necessity for remote monitoring comes to the fore with the following practical user journeys, as showcased in *Figure 11.1*.

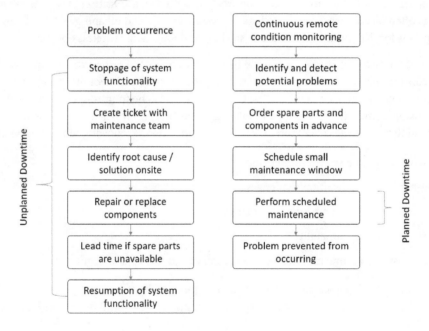

Figure 11.1 – Comparison of user journeys without and with remote monitoring

The column on the left is a system where remote monitoring is not implemented. This is a reactive system, and when a problem occurs, the system functionality is stopped. Teams scramble to assess the issues, understand the root cause with limited to no historical data and patterns, and then suggest a fix. If it involves a replacement of a faulty component, there is a lead time to procure it, after which the system is set in operation. The entire duration of **mean time to repair** (**MTTR**) is very high, creating significant opportunity losses and stress on the maintenance personnel. If the system is in an inaccessible location, for instance, in an oil pipeline midstream through a dense forest, the situation becomes all the more complex.

The column on the right depicts the user journey for a system with remote condition monitoring enabled and powered by IIoT. Continuous availability of data from operation and control systems, in-situ sensors, and auxiliary systems helps create a real-time health view of the asset. Potential failure scenarios and **mean time to failure** (**MTTF**) at the detail of individual components are predicted through data analytics algorithms. Spare parts and service personnel are mobilized well ahead of time. A planned downtime in off-peak hours is selected, and scheduled maintenance takes place. The system is back and running, and the problem is proactively prevented in the first place. The advantages immediately outweigh the upfront investments needed to set up such a system, and the return on investment is rather easily quantifiable.

Drones and robots to the rescue of oil pipelines

Oil pipelines and their infrastructure spread across thousands of kilometers across onshore deserts, swamps, mountains, marshlands, and offshore beneath the ocean. There is a strong need for frequent inspection for maintenance and ensuring safety standards. Static monitoring infrastructure is complex, expensive, and involves disruption to operations. Unmanned aerial vehicles and mobile autonomous robots come to the rescue, scanning hundreds of kilometers of pipelines for anomalies and sabotage and fixing them immediately. Advanced sensors and imaging devices on the drones help identify issues, and the wheeled robots on the ground help by flagging and fixing solvable problems (tightening of valves, etc.).

Since remote monitoring from a central location is highly dependent on the availability of data in real time, it becomes a constraint where data transmission is a hassle. Remote locations lack basic connectivity infrastructure and are severely lacking in terms of the technology backbone to connect to the internet. Speed (lack of), bandwidth, latency, and loss of data packets are some very serious factors that get in the way of delivering cutting-edge remote monitoring solutions. Let us now explore these problems with regard to long-range data acquisition and ideate on steps to mitigate them taking industrial success stories as reference.

Long-range data acquisition challenges

Data acquisition is principally the most important problem in the data processing stream. While there have been tremendous advancements in technologies and consortiums creating interoperable communication mechanisms, applicability, coverage, and scale are still not globally uniform. User adoption is also a key to the popularity of these protocols and vice versa, creating an infinite chicken-and-egg paradigm. The various technologies relevant to long-range data acquisition are plotted in *Figure 11.2*.

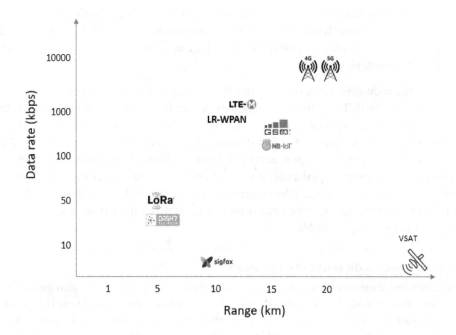

Figure 11.2 – Wireless transmission technologies based on distance and bandwidth

As can be seen, the range of transmission is plotted on the x axis, while the data bandwidth is on the y axis. Speed, bandwidth, and data volume are key parameters for choosing technologies for implementations, especially for remote locations, as they pose a significant architectural dilemma.

There is always a design decision that needs to be made between processing and inferencing at the edge versus the cloud. Each has its own benefits and use cases, and choices need to be based on requirements and constraints while deciding on an architecture. A non-exhaustive list of scenarios where the edge and the cloud come to the fore is tabularized in *Figure 11.3*. Decisions and control involving a real-time response (such as shutting down due to alerts) are usually offloaded to the edge controller.

The edge also takes the role of the first responder, but in autonomous systems, the entire control is vested on the edge. Results and decisions are documented and logged to the cloud to be presented to the stakeholders. Having said this, the edge needs to allow for being securely connected to and controlled by operators. When multiple instances are to be centrally managed and when humans are needed in the decision-making process, the data inference is either communicated to the cloud or computed in the cloud. For instance, the edge monitoring system may prescribe preventive maintenance, but the central monitoring system schedules are based on data from global resources, such as the status of nearby locations, and the availability of resources.

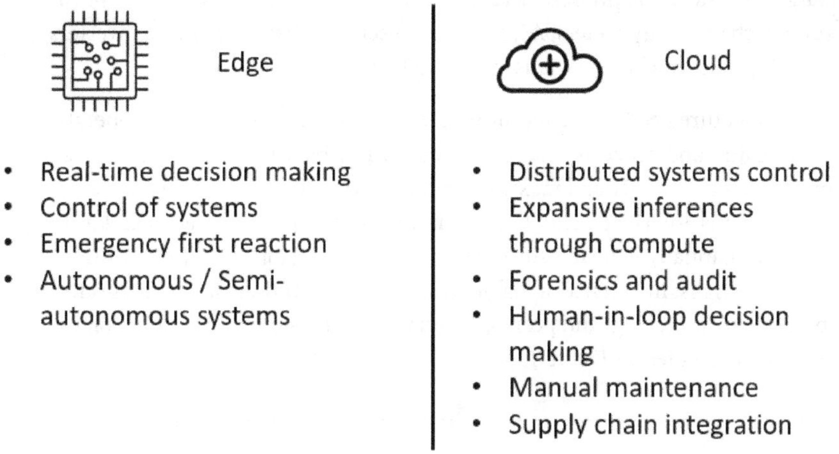

Edge

- Real-time decision making
- Control of systems
- Emergency first reaction
- Autonomous / Semi-autonomous systems

Cloud

- Distributed systems control
- Expansive inferences through compute
- Forensics and audit
- Human-in-loop decision making
- Manual maintenance
- Supply chain integration

Figure 11.3 – Scenarios for offloading decisions at the edge versus the cloud

Adaptive bitrates built-in cameras

IP cameras have evolved and are now capable of running fully trained **machine learning (ML)** models on them, so much so that the cameras have the intelligence to respond to scenarios and adapt their behavior. Using **computer vision (CV)** models at the camera's processor, they are able to identify active areas of interest within their field of vision. They either zoom in on the direction (intrusion, active movements, etc.) or crop the visual frame to carry only relevant data pertaining to the event. Cameras are also able to detect periods of inactivity and switch to low-bitrate mode (reduction in frames per second) in order to conserve transmission bandwidth. They also are capable of sending the bites of a frame that keep changing, and the reconstruction algorithm at the edge deciphers the entire frame based on pilot frames in order to conserve data in constrained environments. A simple camera has now evolved into an intelligent eye inching toward human perfection.

In order to present the principles of architecture in a constrained environment, we undertake a journey through two very interesting use cases. One is a solar farm that captures the sun's renewable energy into electricity, and the other is a carbon capture and storage plant. Both of these are constrained environments in terms of their location, accessibility to human personnel, and the availability of data transmission facilities.

Case study 1: remote monitoring of a solar farm

Solar farms are widespread geographically and help tap incident solar energy into DC electricity. Each solar farm is composed of multiple arrays of solar panels connected in series. These solar panels, in turn, are made of photovoltaic cells. Photovoltaic cells have material characteristics that, when exposed to light, change their electrical properties (current, voltage, and resistance). Solar farms contain other infrastructures such as energy meters, AC isolators, fuseboxes, battery storage, inverters, DC isolators, cabling, mounting, sun-tracking systems, and grid transmission systems.

The entire infrastructure needs to be monitored to prevent sabotage, measure operational efficiency, monitor asset health, and schedule preventive condition-based maintenance. Thus, monitoring is essential for the high availability and maximum productivity of the solar farm. Since the infrastructure is expensive and farms are usually situated in remote areas (owing to large open land requirements), monitoring them manually is nearly impossible due to accessibility and labor limitations. Remote intelligent monitoring is, in essence, a viable business solution that can be integrated with cleaning systems, operations monitoring, and perimeter surveillance systems. The entire solar farm ecosystem at a high level is represented in *Figure 11.4*.

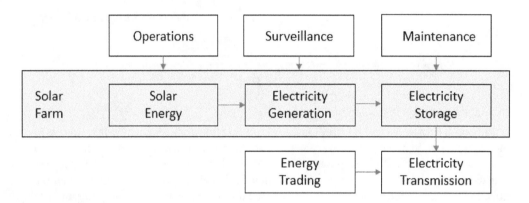

Figure 11.4 – Solar farm structural ecosystem

The obvious advantages of a remote monitoring system are the avoidance of unplanned downtime and expensive on-site checks and maintenance by personnel. As we will also see in the next section, manual maintenance drastically increases the carbon footprint, going against environmental sustainability.

Solar farms are prone to four substantial risks, which the remote monitoring system aims to detect, prevent, and mitigate depending on the level of implementation. This is visually represented in *Figure 11.5*.

Figure 11.5 – Risks and challenges in a solar farm

Let's look at each of the four risks in a little more detail:

- **Fire**: Solar farms generate substantially high amounts of heat by solar thermal capture and high current passthrough. This creates high-resistance joints and component degradation. Non-detection of these heat islands and non-replacement of aged components in time create the risk of fire, damaging the entire installation. A distributed sensor system for continuous monitoring of current, temperature, and humidity can help detect fluctuations, and control measures can be deployed.

- **Sabotage and theft**: As we have seen, the solar infrastructure required to generate, store, and distribute is expensive, and owing to the remoteness of the locations, farms are sometimes subject to theft and sabotage. A robust surveillance system is imperative in this scenario that seamlessly operates 24x7x365. A combination of audio, visual, and thermal imaging helps safeguard the perimeter. Taking bandwidth limitations into consideration, data intelligence must be employed, as seen in the previous section. This essentially involves either offloading inference-making and corrective actions to the edge or employing adaptive data transmission mechanisms to the cloud.

- **High operational expenditures**: Solar farms are usually situated in areas that are arid and receive higher solar incidence. This, in turn, means a lot of dust, which settles on the panels. Dusty panels have drastically reduced energy conversion efficiencies and hence need regular cleaning. The farm itself needs to be gardened frequently to remove weeds and plants from growing on the panels' surface. Also, if the asset health routine is not automated using sensors, it's a labor-intensive process with on-field measurements during the hot sun. The maintenance team also needs to travel to the site adding to the carbon footprint on top of **Operational Expenditure (OpEx)**.

- **Cyberattacks**: Solar farms are also subject to cyberattacks with rogue devices that can enter the physical network. Since most farms are connected to the utility grid, they also increase the attack surface and potentially need to be fortified to prevent any type of untoward cyber threat.

The architecture for such a solution, as we have advocated earlier, is based on requirements and constraints. There should always be the right balance between over-engineering versus building for the future. Most importantly, the architecture needs to be built on open standards and modularized components to make future integrations and new services added as coherent as possible. Let us now explore the solution architecture design of the solar farm.

The solution architecture for remote solar farm monitoring is presented in *Figure 11.6*. The major components of the solution are labeled and described after the figure. A single aggregated monitoring platform could also be designed to serve multiple solar farms in a group. The solution is presented as an edge-heavy solution where the calculations, inferences, and decision-making are offloaded in-situ, making it robust in case of network unavailability.

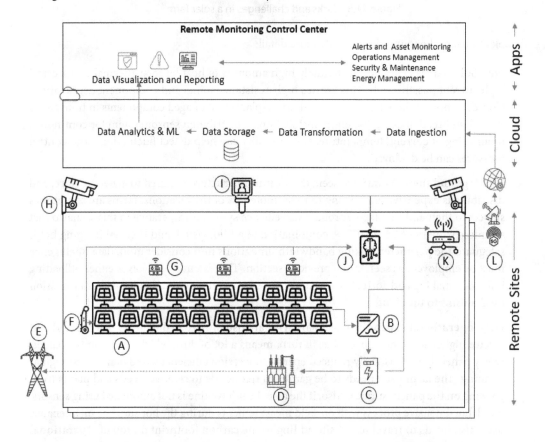

Figure 11.6 – IIoT solution architecture for solar farm remote monitoring

The description of the architecture is broken down into multiple layers for ease of understanding:

A. Solar photovoltaic panels

B. Solar charge converters/controllers

C. Energy conversion (inverters and storage)

D. Transformer

E. Energy grid

F. Autonomous solar panel cleaning robot

G. Sensor unit (temperature, humidity, energy meter, luminous intensity, and proximity)

H. CCTV visual surveillance

I. Thermal imaging camera

J. Edge computing unit

K. Industrial gateway

L. 5G/satellite radio communication

The solar photovoltaic cells are installed across the farms, each having a dedicated charge converter and controller unit. These are then interfaced with inverters and energy storage units on site. To transmit power to the energy grid, transformers are employed to step up the generated voltage. This, in essence, constitutes the functional build-up of a solar farm. Since this is a complex control system, there are thousands of sensors and data processing units in play that seamlessly generate and efficiently control electricity and the system. The next few sections describe in detail the data processing systems at the edge (on the farm) and on the central monitoring system (in the cloud).

The heavy edge configuration

The edge computing system is configured with high compute and storage capabilities and also built-in redundancy in case of a potential hardware failure. The sensing unit is built modularly, consisting of temperature, humidity, luminous intensity, and proximity sensors. If the temperature exceeds the average temperature threshold of the farm, an alert is generated to take corrective action. Corrective action can range from shutting down the solar panels to activating water sprinklers to cool down localized zones.

The installed current sensors also transmit real-time current generation values to the edge, where it computes the average value across the farm with historical values of the individual panel to ascertain conversion efficiency. If the generated output power is less than the ideal output, the edge unit takes corrective actions. Intelligent actions such as activating the autonomous panel cleaning robots are taken after computing data from nearby panels, as there might be dust settlement on the panel interfering with power production. This is a very crucial step to increase power generation efficiency. Imagine a manual cleaning process for the panels every time there is dust on them (which is often the case in arid areas).

In order to enable local surveillance, three strategies are employed. Firstly, a 360-degree rotating network of IP cameras is deployed along the perimeter. The direction and focus are controlled at the edge. The video frames are analyzed in real time for non-desirable movements (intruders, animals, etc.). The thermal imaging camera is also deployed to confirm the same. Local warning systems and speakers are employed to deter such incidents and the central system is notified of the perimeter breach. The recorded video footage is intelligently cropped to optimize bandwidth (for the region of interest, trimmed for the duration of the incident, and frames reduced for the optimized record).

Central cloud-based monitoring platform

The data acquired through the ways stated previously is then sent to the central monitoring platform for users to analyze and take action if needed. The ML model running in the cloud trains on user classification and decision behavior to correlate for future scenarios. This ever-evolving model is then sent to the edge to be deployed for an increasingly improved detection and response management system.

The cloud monitoring system has separate serverless functions to ingest data, filter, and store it in an efficient manner. Separate algorithms are employed on the data to calculate global operating KPIs. This data is then presented to the end users depending on the user roles. For example, the sustainability score and renewable efficiency parameters are presented to the chief environment officer, while parameters such as return on investment, reduction in maintenance costs, and operational efficiency improvements **year on year** (YoY) are all presented to the executive leadership. More technical data on the solar farm performance and opportunities for improvements are presented to the technical engineering stakeholders for them to work on continuous process and product improvement. Energy production data and feed data are all presented to the utility and grid stakeholders and stored for future audit purposes.

Data communication

There are three types of data communication in this scenario:

- Integrating data from already commissioned individual systems such as SCADA, historians, charge control unit software, and energy transformers
- Collecting data from sensors on the field and transmitting them to the local edge computing unit
- Data transfer from the edge to the central cloud monitoring platform

Solar farms are typically integrated as a combination of multiple independent software-hardware systems into an integrated solution. Solar panels and control systems are sourced from different OEMs while charge controllers and batteries are procured from other OEMs. They are then put together and integrated into a single functioning unit by a system integrator. Having a standard mechanism and protocol for data exchange becomes imperative for seamless and comprehensive coverage. The good news is that the edge system in our scenario is open and can be configured to support multiple protocols and integrations. Different data definitions (IEC-61850, IEC 61450-25, and RDS-PP) can be supported.

Sensor-to-edge communication can be handled in multiple mechanisms. LoRaWAN, local Wi-Fi based on IEEE 805.14, and a mesh network based on Sigfox can all be established with relative ease. The choice also depends on the existing ecosystem, number of devices, size of the farm, bandwidth requirement, latency, and data volume.

Edge-to-cloud data transmission is an important topic in this context, as connectivity cannot be assumed to be available all the time. While the penetration of 4G/LTE has expanded the coverage and bandwidth comparatively, it is still not available across all geographies. Satellite communication with constrained data transfer is also an option for extremely critical data transmission (SOS intimating fire accidents, etc.).

Data analytics

Data analytics and ML create a plethora of applications that were difficult or non-existent before. Energy production prediction based on weather patterns, historical production data, and current operational efficiency patterns is a great use case, to begin with. There are other use cases, such as predictive maintenance, and solar panel tilting for maximum irradiance that can be developed as the solution matures. Thus, data (as has been emphasized repeatedly) is the fuel that powers innovation and futuristic and advanced application development. The completeness, context, and timeliness of the data which forms the metadata are equally if not more important in order to successfully generate inferences.

Concluding remarks

The edge, in this case, has higher autonomous decision-making powers built in order for the system to operate even during network outages. But a cloud-based central monitoring system is imperative to drive a comprehensive operating solution. As discussed in the earlier section, to optimize the available bandwidth, the edge can perform local processing of data while uploading inferences, alerts, decisions, outliers, and performance KPI values to the central monitoring system. By correlating this data from the entire ecosystem, alerts can be analyzed in depth to arrive at decisions regarding maintenance plans.

The maintenance is thus changed from timely scheduled maintenance to dynamic condition-based maintenance. Irreversibly damaged panels and components nearing their end of life can be identified beforehand with pinpoint accuracy and can be sourced from suppliers just in time for the maintenance activity. With a lean supply chain (inventory) and an effective maintenance plan, the whole operation is now made more efficient.

In order to create a platform-based architecture, we present an AWS-based reference in the next section. Both edge and cloud services from AWS are leveraged to create a scalable service-oriented framework for solution building.

AWS reference architecture for solar farm remote monitoring

The AWS reference architecture for the solar farm remote monitoring is presented in *Figure 11.7*.

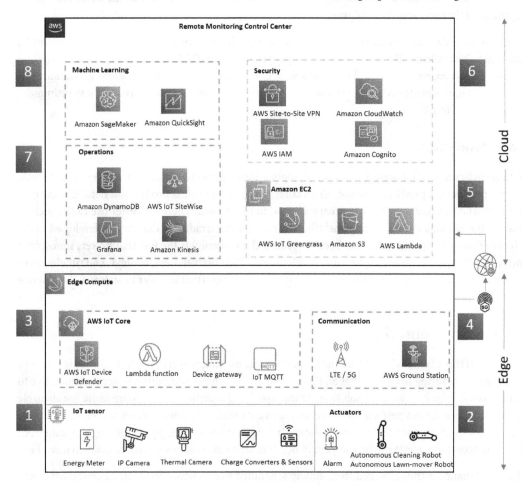

Figure 11.7 – AWS reference architecture for solar farm remote monitoring

Let us look at each of the layers, numbered 1 through 8:

1. **IoT sensor:** The IoT sensor layer consists of in-situ sensors that are deployed for the measurement of physical parameters (temperature, current, power, power factor, humidity, light intensity, etc.).

2. **Actuators:** The actuators consist of systems that are controlled by remote monitoring systems. These can include third-party systems with communicable interfaces. For example, smart water sprinklers, autonomous cleaning robots for panel cleaning, autonomous lawn mowers for field clearance, alarms, and loudspeakers belong to this layer.

3. **AWS IoT Core**: The AWS IoT Core is the edge compute layer responsible for decision-making at the edge. This has built-in connectors such as OPC UA and MQTT for data acquisition from existing systems such as SCADA and energy management systems. This also has built-in lambda functions for data acquisition, aggregation, and inference formation. IoT Defender is built to provide an important layer of device security at the edge.

4. **Communication**: Data transmission via the gateway to the cloud happens either via the 4G/5G antenna or through a satellite ground station service. Intelligent routing decides the channel and the payload depending on criticality.

5. **Amazon EC2**: The **Elastic Cloud Compute** (**EC2**) service in the AWS cloud comprises the mirrored central version of AWS Greengrass, `lambda` compute functions, and S3 storage services. Data ingestion, preparation, and transformation happen in this layer.

6. **Security**: Standard security services such as identity and access management and CloudWatch enable the security of the system. A secure site-to-site VPN service is deployed to obtain and send secure information to the edge and back.

7. **Operations**: Metrics and KPIs are streamed in this layer and Grafana dashboards are fired up to visualize key performance indicators and alert when outliers are detected.

8. **Machine Learning**: The AWS SageMaker service is used to run complex ML algorithms. The inferences are visualized on a QuickSight dashboard.

We now traverse onto a topic of great interest in today's context. As seen in *Chapter 9, Taking It Up a Notch – Scalable, Robust, and Secure Architectures*, the three pillars shaping global digital ecosystems are autonomy, interoperability, and sustainability. The next section is an important section that focuses on the topic of sustainability. Sustainability is at the core of design and decision-making for many enterprises that wish to reduce the impact of globalized industrialization. This is a great and important measure as we reach for the creation of a conducive environment for future generations. Architecture is the fulcrum of this topic, as will be clear as you progress through it.

Architecting for sustainability

From an overall success point of view, a smart IIoT solution is one that is desirable from an end user perspective, viable from a business perspective, and feasible from a technology implementation perspective. A fourth, and probably the most important, pillar is whether the architected solution is sustainable from an **environment, social, and governance** (**ESG**) perspective. This is described in *Figure 11.8*.

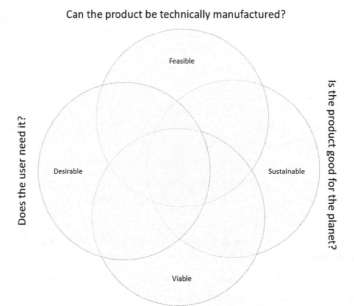

Figure 11.8 – The magic intersection of solution design

The intersection between the four circles is the magic or sweet spot for a great architecture design. The idea, mathematically, would be to make all the circles coincide with each other, which in practice would be far from ideal.

A product or solution architected following the principles of desirability, usability, viability, and sustainability and manufactured and implemented with digitalization at the core creates a path-breaking business model – shifting tangentially from product-based to service-based. This concept fuels the basis for a circular economy. Circular economy refers to a production-consumption model that comprises time-sharing, reusing, repairing, and refurbishing existing materials and products over as many cycles as possible until the useful life limit is reached.

The UN World Commission on Environment and Development defines sustainability as "*development that meets the needs of the present without compromising the ability of the future generations to meet their needs.*" The term ESG forms an important subset of sustainable design, but we will focus on the environmental aspect for the scope of this book, as depicted in *Figure 11.9*. The social pillar, a fundamental building block, primarily consists of topics related to **Diversity, Equity, and Inclusiveness (DEI)**, which is how employees are treated and engaged. Governance deals with company policies, ethics, and values and is more of a philosophical one. Readers are encouraged to explore these vital topics on their own to get a holistic picture of macro policies that influence micro-system design.

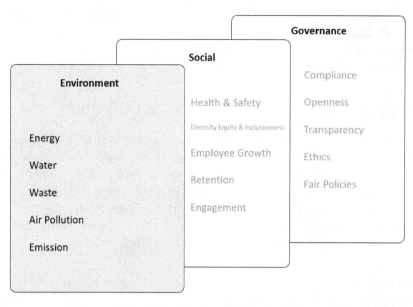

Figure 11.9 – ESG with a focus on the environment

The environment block of ESG aims to minimize the impact on the environment by initially measuring, then monitoring, and then controlling. A carbon footprint report or a sustainability report captures the performance of the enterprise in accordance with these parameters and the corrective actions taken to progressively minimize the impact on the environment can be used as a benchmark. This consists of topics such as energy, water resources, reduction in wastage, air pollution, and emissions:

- **Energy**: Reliance on renewable sources of energy. Becoming independent of the grid by utilizing on-site energy generators and giving back excess energy to the utility grid.

- **Water**: Minimizing water consumption as part of the process and maximizing alternatives to water. Net-zero wastewater discharge outside or to water bodies without proper treatment and removal of harmful solvents. Integrated effluent and sewage treatment with rigorous monitoring and quality control.

- **Waste**: Reduction in wastage emitted out from the processes. Emphasis is on recycling by-products and non-useful waste into applicable and useful products in the value chain and may be across verticals.

- **Air pollution**: Constant measurement of air quality of the facility with stringent measures of quality control. Alternate measures to minimize polluted air being released into the atmosphere (chemical adsorption, etc.).

- **Emission**: Efficient carbon and harmful gas capture and conversion facilities.

With the ideologies of sustainability deepened, let us now embark on the second case study of the chapter, specifically aligned with this topic: **Carbon Capture and Storage (CCS)** systems. This is a very interesting topic in today's context, and with the evolution of technology (IIoT and communication), it becomes possible to architect and design solutions for this problem. Let us venture forward.

Case study 2: remote monitoring of a CCS unit

So, what is CCS, and why is it important to enterprises and countries across the globe? Carbon capture is the process of encapsulating carbon dioxide, which gets emitted by industries during the production process. CCS is used by heavy industries (cement, steel, oil, plastic, machine building, and power plants) to reduce their greenhouse gas emissions, helping them toward achieving net-zero emission scores. The captured carbon dioxide is then stored far below the earth's surface. There are many interesting applications for how stored carbon can be utilized. Globally there are around 100 CCS projects live, and they play a role in offsetting global warming.

> **The origin of CCS**
>
> The initial idea of capturing carbon dioxide and preventing it from being released into the atmosphere originated in the 1970s. At a gas processing unit in Texas, USA, the captured carbon dioxide was released into a nearby oil field. As a result of this, the crude oil that was obtained from the well was enhanced. This process, known as enhanced oil recovery, has proven super successful for oil companies. By capturing CO2 from the exhaust and repumping back, the fossil fuel was enriched, resulting in an almost circular economy model.

At a high level, CCS or **Carbon Capture Utilization and Storage (CCUS)** consists of four processes: carbon capture, transportation, sequestration, and utilization. Each is a critical process and monitoring their parameters and operations is critical. Gas leakage at any point can prove counterproductive. The high-level sequence of processes is showcased in *Figure 11.10*.

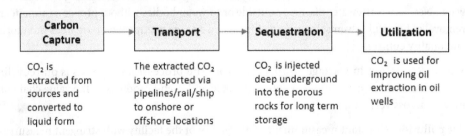

Figure 11.10 – Carbon capture storage and utilization process

The exhaust gases emitted from power plants or cement or steel plants, for instance, are run through a solvent. Carbon dioxide from the emission gets absorbed and attached to the solvent, and the rest of the gases are returned. Then the solvent is heated up to liberate the carbon dioxide molecules.

These are then cooled and ready to be transported to carbon wells. Carbon wells are deep pits below the surface of the earth and are located on land or offshore near oil wells. Dedicated pipelines, rail freight, trucks, and container ships are used for transportation. At the receiving plant, carbon dioxide is further heated to 31 degrees Celsius and with the application of high pressure, the compound is ready to be pumped below the earth's surface, where it readily combines with underground water, carbonating it. This interesting journey of the carbon molecules from the exhaust to the earth is shown diagrammatically in *Figure 11.11*.

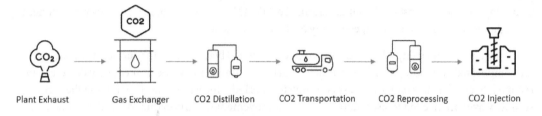

Figure 11.11 – Carbon dioxide from the exhaust to deep earth

The efficacy and the utility of such a system are still debatable, and we will let the readers explore and draw their own conclusions. But one thing is very clear, the need to monitor the entire operations and the extremely remote nature of the setup. Hence, we have chosen this topic as a case study for this chapter. For the sake of simplicity, since the utilization applications are varied, we limit the use case to carbon capture and storage.

The solution architecture for the CCS unit is presented in *Figure 11.12*.

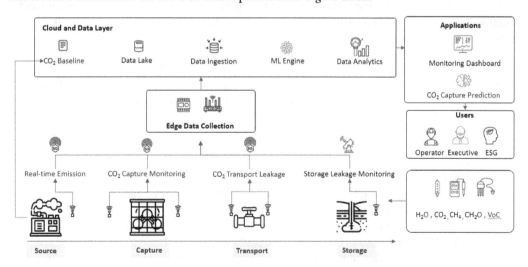

Figure 11.12 – IIoT solution architecture for CCS unit

Continuous and automatic monitoring systems are essential across the entire process for meeting compliance regulations. Reporting the data to a central agency is also needed at regular intervals for process audits. At each stage of carbon capture, transportation, and carbon sequestration (storage), sensors and measurement systems are employed. These automatically measure the presence of up to 16 compounds, such as carbon dioxide, water, methane, volatile organic compounds, carbon monoxide, nitrogen dioxide, and ammonia. The sensor units are usually placed as a network in remote locations. For example, at the exhaust outlet of a power plant and the exit point of the carbon dioxide distillation plant. The sensor data is transmitted to the edge processing unit via either 4G LTE/5G or WAN technologies based on installed configurations. LoRaWAN is a very suitable technology component for high-speed local data transfer, especially at the capture facilities.

At the sequestration facility where carbon dioxide is pumped below the earth's surface, measurements are complicated and more crucial. Any leakage during the transfer must be immediately detected and rectified. Even after the storage process is completed, periodic measurements by studying the tectonic seismic waves are needed to ascertain leakage and movement of carbon dioxide in the stored earth faults. Fiber optics-based leakage monitoring systems are also prevalent.

Due to the extreme remoteness of either on-field wells or offshore wells, manual measurements are extremely difficult. Thus, automated remote monitoring and data transmission using suitable mechanisms are needed for successful solution implementation. Satellite communications and **Maritime IoT (MIoT)** are usually recommended as primary data transmission channels with a fallback on mobile wireless telephony subject to availability.

The central management platform collects data from multiple sources of the process, cleanses data, and helps create a unified data model. This model is then compared with a baseline model to detect differences. Alerts are also visualized on the central management console, and corrective actions and preventive measures are undertaken. The central platform on the cloud also serves as a repository for compliance and legal audits in the future.

These two use cases have, in our opinion, provided significant coverage in terms of the challenges of establishing scalable architectures for remote monitoring applications and ways and methods to overcome them. As always, there is no single solution that fits all equations and the design needs to be adapted based on needs and evolution.

Low orbit pico-satellites

A lot of interest has been garnered around the launch of multiple low-orbit miniature pico-satellites capable of providing low-bandwidth connectivity from any corner of the globe at a very attractive and affordable price point. These satellites orbit at 450-550 km altitude, providing global coverage and enabling IoT devices to operate from literally anywhere. These satellites could be groundbreaking for enabling remote monitoring in inaccessible regions. The footprint of the satellites is minimal, and they are self-sustaining in terms of energy and operations. Integration with a wireless carrier technology would provide a robust ecosystem for end-to-end data coverage. The future is brighter!

Summary

This chapter was the penultimate of this book and focused on practical challenges in implementing remote IIoT solutions. Remote monitoring is one of the low-hanging applications of IIoT, where the benefits clearly outweigh the investments made. Remote monitoring also brings about the opportunity to move toward service-oriented architecture design, which we hope was clear throughout the chapter. The number of applications and business use cases that can be constructed and integrated with such a platform is large.

After the case for remote monitoring was presented and cemented, we jumped into the challenges in data acquisition and building a solution architecture around the mitigation path.

Sustainability was one of the key messages from this chapter and will be very prominent in the coming years. Designing solutions around constraints and requirements focusing on value drivers, especially sustainability, is going to be the key as we build solutions for the next generations to consume. The core concept was presented using two solid use cases. The first one was around remote solar farm monitoring. The AWS reference architecture for this solution was also presented. The second use case was around architecting a solution for carbon capture and storage plants.

This chapter delved in detail into energy conservation as a topic as part of ESG and sustainability. The next chapter is a hands-on chapter and will deal with the application of data science to energy datasets. These are super helpful examples that will help you understand and assimilate the process for orchestrating ML-based applications. Once equipped with the knowledge, methods, and tools, the journey to solve the next big problem is chartered. And by the way, it's the last chapter of the book.

12
Advanced Analytics and Machine Learning

Hopefully, if you have gotten to this point on your journey, you are excited about the possibilities of getting started. Getting to this point has been a workout, but our ultimate goal is to thoroughly understand the how and why of good architectural decision-making. **Machine Learning** (**ML**) and **Artificial Intelligence** (**AI**) is an exciting topic that is still relatively new in our industry. Performing inference on data in real time to determine a course of action is powerful. Performing inference on your data is often a phrase you will hear in ML. It is the act of reaching a conclusion based on evidence and reasoning. In our case, we will look at how to infer a result or condition based on previous or existing data.

The authors of this book are not data scientists. Generally, we are IoT and cloud architects with a broad set of skills focused on the topics needed to deliver a complex IoT solution, including experience in data collection, engineering, and collaborating with analysts and data scientists to drive complex solutions. We want to clarify that this chapter will not make you a data scientist; instead, it will show you how to incorporate and work with data scientists in making better-informed decisions about your environment. Additionally, we will illustrate the end-to-end process from data consumption to analysis. This approach is probably very similar to your goals.

Our focus for this chapter is to provide a contrived ML example and derive a model based on collected sample data, and we spend a lot of time just getting the data. You can build on these examples across the entire book to provide data in a manner that is useful for analysis, and then once a model is determined, you can then build it into your overall process. In this chapter, we are going to cover the following main topics:

- ML and industrial IoT
- Engineering your data
- Building the model
- Real-time inference

Technical requirements

The technical requirements for this chapter are similar to previous hands-on technical chapters. A minimal technical background is required if you want to read along and understand the process. However, working alongside the examples and recreating the scenario requires additional technical skills. Working through previous chapters should be beneficial to get some necessary experience with AWS IoT Core, AWS Glue, and AWS IoT Greengrass. We dive deeper into getting better Modbus data and expand our AWS Lambda examples to decode that data. Knowledge of Python programming is a plus for lambda work. It will also be beneficial because we will dive deeply into analysis and model building using Python with a Jupyter notebook.

You can find the code samples mentioned in this chapter at `https://github.com/PacktPublishing/Industrial-IoT-for-Architects-and-Engineers/tree/main/chapter12`.

ML and industrial IoT

AI and *ML* are such overloaded terms at the moment. Typically, when something looks new and exciting, the vendor marketing machines get ahold of it and turn it into anything and everything they can sell. There is no doubt that advances in AI and ML in the last few years have been tremendous, and bringing this capability to the masses, or at least to the technical community, has been one of the more significant focuses of cloud providers. Initially, we will outline the concepts of AI and ML and steer the focus of this chapter toward our final goals.

AI is a grand concept focused on creating intelligent or smart machines; its goal is to mimic human-type intelligence, specifically the ability to learn and solve problems. Additionally, there is a recent trend toward refocusing the definition of AI to encompass the idea of rationality, meaning machines that can act rationally in their decision-making. This concept provides a broader viewpoint that allows AI to deviate from comparisons with human intelligence and decision-making into something intelligent but not necessarily human. AI is divided into several areas, such as natural language processing, vision or image recognition, and ML, each focusing on advancing the field in a specific way.

As mentioned, ML is a significant subset of AI and the primary focus of this chapter. ML aims to create systems that can learn and adapt based on the patterns within existing data. We can then analyze data and derive relevant patterns encapsulated in models using applied algorithms and statistical models. These pre-built models can draw conclusions through interference and provide decision-making capability based on incoming data. This result is precisely our goal for this chapter – not to draw specific conclusions but to show how a model can be leveraged to draw such conclusions once it is created.

Interference versus rules

When setting up an IoT environment, an initial approach is often to use rules and alerts to identify and manage potential problems with your system. Generally, this approach is okay. If you know the temperature should be between specific values, you can check for that range and trigger a warning when the value does not match. This is a clear example of rule-based monitoring: a simple rule defined to check, say, a temperature value. Is the temperature between 30 and 100 degrees Fahrenheit? If so, perform some action.

More complex rules can also be analyzed, such as *if* the temperature is in range *and* the pressure is below 30 PSI, then do nothing. It is easy from here to conjure all types of situations using equals, greater than, less than, and so on. These rules and ranges can be derived from the equipment service manual or the current operators and maintenance team.

AWS IoT rules are a simple rules engine that can route messages based on type or content. You will see an example later in this chapter. However, messages are often encoded or encrypted when sent to IoT Core. As we have seen, encoding is used to reduce the bandwidth of a message, either over the air or along the wire, making messages smaller and quicker to send.

Consider our Modbus data packet in the following code snippet. We can easily pick out and route the message based on the value of `type` or `id`; however, the value is provided as two 8-bit signed integers. We need to convert these values into a single 16-bit unsigned integer before deciding on the data. This makes it harder to use a simple out-of-the-box rules approach on some data:

```
modbus/response/conveyer
{
    "type": "ReadInputRegisters",
    "id": "00824502",
    "bytes": [
        0,
        15
    ]
}
```

Even the most complex custom-built rules engine cannot or should not be required to decode before evaluating the measurement. Therefore, data should be sent to some intermediate asynchronous service for evaluation. *Figure 12.1* outlines a sample flow for the overall data engineering process. Note the addition of data flow to a separate rules engine. Ideally, this would be a data queue where data can be queued and evaluated quickly but without impeding data flow through the central system.

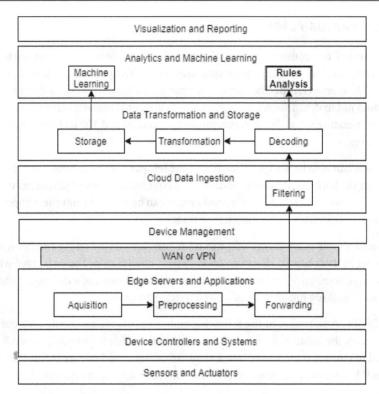

Figure 12.1 – Data flow and rules processing

This section's title is *Interference versus rules*, but there is no reason why these two do not always work together. Some rules can be considered very rigid. Spoilage can occur if cold storage is not within a defined range of degrees. This type of monitoring would be overkill for using ML, so rules can and should be an integral part of your monitoring environment.

Learning from the data

I find it interesting that often we don't know what *normal* means with our equipment. Operators can often tell when something is wrong or something will soon go wrong, but they may have trouble explaining it clearly. IoT focuses on using data to bridge that gap in understanding – knowing the parameters for normal operating conditions versus data that is out of range and unexamined.

ML is used when we need to go beyond rules, even complex ones, to define known parameters or conditions for our equipment. ML is a subset of AI that allows us to examine data in new ways and build models that can learn from existing data and be improved over time, to predict anomalies or failure of equipment or processes.

Engineering your data

Our overall goal is to combine some of what we have learned from previous chapters – gathering data from the edge, building on what we learned in *Chapter 10, Intelligent Systems at the Edge*, and processing and storing data that we covered in *Chapter 8*. We will configure the addition of ML to this overall process and then use the model we will build for real-time inference of new incoming data.

Figure 12.2 outlines our target architecture for this example, so let's walk through the areas on which we will focus:

- We start with the modify Modbus TCP example from *Chapter 10, Intelligent Systems at the Edge*, to pull voltage and current data from our simulated PLC. In our last example, we were pulling coil data from the PLC. These are Boolean values signifying the on and off states of a machine. For our use with ML, it is better to pull and use actual decimal values from the equipment. With this example, current data will be queried and examined at 60-second intervals.

- Process and decode Modbus messages. The new values from our simulated PLC require some additional decoding. This is similar to the decoding we did in *Chapter 4, Real-World Environmental Monitoring*, with the encoded LoRaWAN messages. We will walk through a part of the `lambda` function for decoding and storing incoming messages.

- Perform ETL using AWS Glue to prep messages for Amazon SageMaker. Before using SageMaker to build a model, we will perform some ETL using AWS Glue to transform our data into the formatted data lake as a large CSV file. We can then use that new format to ingest into SageMaker Studio.

There is much to cover as prep work for building our ML case. Still, seeing this complete scenario and reviewing some data engineering is essential. The overall flow will help you to understand how the data flows from end to end within the architecture.

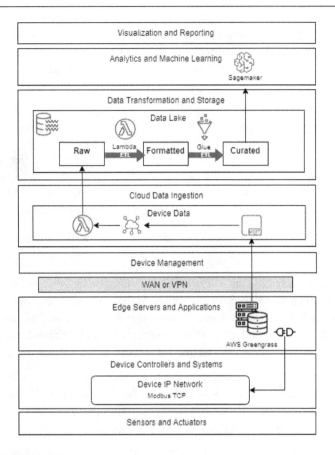

Figure 12.2 – Target ML architecture

The only new component in the preceding diagram is Amazon SageMaker. The data flows have been discussed in previous chapters and should not be drastically new. In this chapter, we expand the architecture to add additional capabilities, such as receiving data through AWS Greengrass or using AWS Glue to perform other ETL.

Retrieving voltage and current data

In *Chapter 10, Intelligent Systems at the Edge*, we built several AWS Greengrass components to interact with our Modbus slave. The Modbus Request component was straightforward in design. It worked by putting a message on the appropriate topic via Greengrass IPC. The Modbus TCP component defines the topic with the structure modbus/request/conveyer. By changing the details within the actual message, we can use a different function and address to receive the needed data. For example, we can modify the message as follows:

```
message = '{ "id": "00824502", "function":
 "ReadInputRegisters", "address": 11, "quantity": 1 }'
```

This message has several changes from what we created in *Chapter 10, Intelligent Systems at the Edge*. We replaced the `id` value with a serial code. This allows us to identify which machine or piece of equipment this request is associated with. We also changed the function from `ReadCoils` to `ReadInputRegisters`. This tells the Modbus TCP component what approach to use when reading data from the PLC. Our `address` space is `11`, not a standard address based on the Modbus specification. Still, different PLCs or Modbus devices vary in address space and the use of absolute versus offset addressing schemes, so we took this example. It will probably take some trial and error for you to get a suitable addressing scheme for testing and when you move things into a more production-focused environment.

The resulting data from this request differs from what we received when requesting coil data. You can see the result displayed in *Figure 12.3*. Since we only asked for one address, the resulting value is two 8-bit signed integers, which need to be converted to a 16-bit unsigned integer to see the actual value. If we were to ask for additional values – say, three data elements – it would return a set of integers corresponding to the initial address space and the two spaces immediately after that one:

```
"bytes": [0,-28,0,-31,0,-25]
```

Once we learn how to decode these values, it should not matter how many we request. We are sticking with one register for simplicity in the rest of the chapter. *Figure 12.3* shows the complete request and response where we only ask for a single register of data.

Figure 12.3 – ReadInputRegisters request and response

This screenshot shows the initial request for data on the `modbus/request/conveyer` topic, and then the resulting data on the `modbus/response/conveyer` topic. Follow *Chapter 10, Intelligent Systems at the Edge*, and update the original query to the one provided in this chapter. You should get this modified response, as shown in the preceding figure. Depending on your PLC or simulator, you may need to modify the address you request. A simulator is the easiest to work with, as you can change values at will and identify which address space you are targeting.

Decoding the values

To decode and process data, we are going to focus on using a `lambda` function. This is similar to what we did in *Chapter 4, Real-World Environmental Monitoring*, when we were dealing with LoRaWAN data that needed to be stored, decoded, and processed for use. We debated slightly about which was the best approach to use for this example. *Chapter 8, Asset and Condition Monitoring*, walks you through the use of AWS IoT Analytics to run data through a configurable pipeline of activities. In the end, we still needed to build a custom `lambda` function to decode the values. That makes our architecture decision easier now, so we can use IoT rules and push the data to a simple `lambda` function for processing. Let's walk through the `lambda` decoding function to see how it works:

```
import json, csv, boto3, os, io, sys, uuid, datetime, dateutil.
parser, time
from botocore.exceptions import ClientError
```

The import and definitions are fairly simple. We will push data to S3 as we have before, storing both the raw incoming message and the transformed message in the formatted data store:

```
def lambda_handler(event, context):
    print(os.environ)
    print("Received event: ", type(event), json.dumps(event))

    eui = event['id']
    print("eui: ", type(eui), eui)

    datenow = datetime.datetime.now()
    print("datenow: ", type(datenow), datenow)
    event['datetime'] = datenow.strftime("%Y-%m-%dT%H:%M:%SZ")
    year = datenow.strftime("%Y")
    month = datenow.strftime("%m")
    day = datenow.strftime("%d")
    minute = datenow.strftime("%M")
    hour = datenow.strftime("%H")
    second = datenow.strftime("%S")
```

The following section is also simple. We print out our lambda environment and the incoming event message primarily for debugging as we build and test the function. We also extract the device EUI, or ID, from the message and break down the date of the incoming message to use all this as partitions when storing and retrieving specific data. EUI is a common abbreviation for the extended unique identifier of a device. We also generate a timestamp and use that as partitions or folders in S3. Ideally, we would extract the DateTime stamp from the message; however, in this case, date and time information is not included in the message. We will have to add it specifically within the function:

```
### store initial message into raw datalake
s3 = boto3.client("s3")
bucket_name = "s3-datalake-iot-raw"
file_name = str(minute) + ":" + str(second) + ".json"
folder_path = "modbus/conveyer/" + str(eui) + "/current/" +
str(year) + "/" + str(month) + "/" + str(day) + "/" + str(hour)
+ "/" + file_name
s3.put_object(Bucket=bucket_name, Key=folder_path,
Body=json.dumps(event, indent=2))
```

Next, we record the raw message in the raw data lake. We discussed this in detail in *Chapter 2, Anatomy of an IoT Architecture*, when reviewing the architectural components of your environment. Note that the folder path uses a strict prefix and then a more flexible folder structure that includes the EUI of the equipment and then the year, month, and day. The minutes and seconds part of the DateTime stamp is used as part of the filename itself within the folder structure.

Now, let's convert our byte string to a decimal value that we can work with later for analysis. We are taking a few extra steps in this process to explain how converting these values occurs:

```
values = event['bytes']
print("bytes: ", type(values), values)
```

The byte passed via our JSON event object is given as a list of signed integers. In this case, we can take the example [0, -6] to represent our incoming value. This statement shows the following output on the console:

```
bytes:   <class 'list'> [0, -6]
```

We include some sample output from the function to see the values and types of different variables during this process:

```
# extract the two signed integers
current1 = event['bytes'][0]
current2 = event['bytes'][1]
```

We extract the two integers from the combined list by referencing their location explicitly. This provides two separate integer variables with the 0 and - 6 values respectively, shown here on the output console:

```
current1:  <class 'int'> 0
current2:  <class 'int'> -6
```

The values provided to us are in a big-endian sequence. This implies that the first byte contains the high-order bits and the most significant value. The second number is the low-order bits in the series.

> **How do you like your eggs?**
>
> I was reminded of this when working out this decoding sequence. The terms *big-endian* and *little-endian* come from the story of *Gulliver's Travels*. For those that have read the book or perhaps seen a cartoon or movie, you will remember that the Big Endians and the Little Endians were engaged in a great war against each other. The focus of the war was based on how they broke their eggs (I assume soft-boiled in both cases), either at the little end or the big end.

We can now convert the signed integers into their hex equivalent values for further processing:

```
# convert the integers to hex strings
current_str1 = hex((current1) & 0xFF)
current_str2 = hex((current2) & 0xFF)
```

We convert the two signed integers into their hexadecimal values – hex strings, to be exact. This is a more expected value for us to work with, and the data is starting to look more interesting:

```
current_str1:  <class 'str'> 0x0
current_str2:  <class 'str'> 0xfa
```

From this point, we can look at the data as actual integer values converting back and forth between hex and decimal equivalents:

```
# convert back to hex integers
current_int1 = int(current_str1, 16)
current_int2 = int(current_str2, 16)
```

In this case, it becomes pretty obvious what our final value will be, but we cannot take this for granted:

```
current_int1:  <class 'int'> 0
current_int2:  <class 'int'> 250
```

The two numbers must be combined to determine the final decimal value:

```
    # shift the first value over 8 places and OR (combine) the
  two values into a single 16 bit integer
    decimal_value = hex((current_int1<<8) | current_int2)
```

This part is a little tricky if you are unfamiliar with bit shifting. Since the first byte is the most significant, we shift it left by 8 bits. This will pad the first-byte value with eight additional zeros at the end. We can use OR between the two values to merge them and fill in the beginning zeros with the values from the second integer. The result is a single combined value:

```
 decimal_value:   <class 'str'> 0xfa
```

In this case, the first byte is zero, so we get a long string of additional zeros. The second byte has the actual value we care about, so we get something such as 00000250 when we combine or use OR on the two values. Okay, it's not surprising here that we get the 250 value in hex as our final result, but try this with other values, and it will make better sense why we have to go through these steps:

```
    current = int(decimal_value, 16)
```

The final value comes as UINT 16 or an **Unsigned INT**eger of 16 bits. This value represents our final current value and what a data scientist or ML models will use to understand how the equipment works:

```
 final current:   <class 'int'> 250
```

We should also point out that we know precisely what our byte list represents within our system. This message and the string are clearly defined to represent current values from a specific piece of equipment, based on the ID or EUI of the machine. This information will probably require a lookup in one of your reference systems or tables. This may occur within the lambda function to determine exactly how you should decode and reference your incoming measurement values:

```
    event['current'] = int(decimal_value, 16)

  # remove unnecessary fields from the json string
    del event['bytes']
    del event['type']

    print("final json: ", type(event), event)
```

Finally, let's clean up our event JSON. We need to add the converted value to the JSON by adding a `current` type to the JSON. Also, we can get rid of some unnecessary values, such as the old byte string and the type element:

```
final json (event):  <class 'dict'> {'id': 'ReadCurrent',
'datetime': '2022-10-07T07:50:24Z', 'current': 250}
```

The final JSON looks clean and readable and is ready for action. Similar to what we did earlier in the function with the raw data store, we can store the formatted final JSON in our formatted data lake. The structure is the same in this case. However, the message is decoded and ready for interpretation:

```
    ### store transformed message into formatted data lake
    bucket_name = "s3-datalake-iot-formatted"
    file_name = str(minute) + ":" + str(second) + ".json"
    folder_path = "modbus/conveyer/" + str(eui) + "/current/" +
str(year) + "/" + str(month) + "/" + str(day) + "/" + str(hour)
+ "/" + file_name
    s3.put_object(Bucket=bucket_name, Key=folder_path,
Body=json.dumps(event))
```

In earlier chapters, we showed how to enhance the message, which you would surely want to do to make data more usable. For our purposes, this is more than enough for an example of how to enhance your data. We want to get a baseline of existing data to build some ML models:

```
    ### placeholder for model integration ###
```

Also, add this placeholder to your code. Here, we are putting a placeholder into the function to use later in our example. After we build our model, we can reference it here to do some on-the-fly anomaly detection:

```
    ### finish
    return event
```

We have already shown most of this in earlier chapters. This was an update from *Chapter 10, Intelligent Systems at the Edge*, with more detail from various other chapters. However, this was a new twist by showing one approach to decoding our messages from the Modbus response data. Hopefully, that addition reinforced your learning in handling and processing IoT data.

Building the model

Before we get into the heart of using Amazon SageMaker to develop the ML model, we have a little more data engineering to consider. SageMaker contains a good number of built-in algorithms and several pre-trained models – one of which we will use in the example. The **Random Cut Forest** (**RCF**) algorithm is an unsupervised learning algorithm that detects anomalies in data points from within a set – that is, data points that diverge from a well-structured data series.

RCF is a good algorithm for looking at time series data and determining spikes in data, or possibly some latency or spikes in a dataset due to production or seasonal issues. Because our current raw data is pretty well structured, assuming the value from our simulator is constant or within slight variations, RCF can analyze this data and determine when data points are outside the given target.

A note about architecture and data science

Data science is a growing and complex field. I think it's clear we could only introduce you to some of the data science concepts in this chapter. It is the architect's job to retrieve data from the field and perform some simple data processing work. Data can be curated, processed, and then provided in a separate data lake or data format for the analyst or data scientist to begin their process.

We chose to use an RCF algorithm as an example in this chapter because it is a relatively simple unsupervised algorithm for anomaly detection. This allows us to provide a working example based on data we can collect from within our environment. There are so many additional considerations in selecting a suitable model. Is your data too small to provide the desired randomness? Are you overfitting or underfitting your model? These questions are best left to an ML or data science book for a complete discussion.

RCF has only a couple of parameters available for configuration. You can choose the number of trees the model defines and the number of random samples per tree. You cannot control which feature goes into a specific tree. Because there are only a few parameters, RCF models generally have higher bias and lower variance, which means the following:

- Bias is the difference between the average prediction of the model versus the correct value prediction – that is, the difference between the model's predicted value and the actual value

- Variance is how variable the prediction for a given value is and the amount of change between different datasets

Bias and variance are inversely connected, so it is impossible to have both low or high; the goal is to find a good balance between the two for your model. With RCF, we aim for better accuracy and less bias in testing data. Next, we will look at the steps we need to build and test our model. These include the following:

- Transforming training data

- Setting up and training our model

- Deploying and testing the model

Together, they will guide us through the end-to-end process of building and using the model in a real-world scenario. Moving forward, we need to prepare the data for input into the model. We can process data in several ways. In the next section, we will go back to our earlier example and use AWS Glue to do some simple ETL on the incoming data. Essentially, we need to take all the individual processed JSON files and convert them into a larger CSV file that can be ingested and analyzed by Amazon SageMaker.

Transforming training data

We need to transform and format our data for easier ingestion into SageMaker to train a model. SageMaker has several standard data formats for its built-in algorithms. You can *stream data directly from S3 using Pipe mode*, which saves disc space on your training instances. These instances are launched within SageMaker code to run the training data and build a model.

Our example will provide our training data in CSV format. We are collecting the individual JSON output files and merging them into a single CSV file. We want to offer a simplified CSV with no header and ensure that the target value is the first on the list. So, we need to be sure that the value we are building our model on – in this case, the current measurement – is the first value in our CSV file.

AWS Glue DataBrew has only been generally available for a couple of years at the time of writing. It's one of many tools AWS has available to clean up and format your data. SageMaker Studio has a Data Wrangler component, providing good built-in functionality for pre-processing data. However, DataBrew seems easier for this type of work and can be more cost-effective from a processing standpoint.

For our purposes, AWS Glue DataBrew does the job simply and efficiently. It also allows us to manipulate data as little or as much as required. *Figure 12.4* shows the final completed cycle.

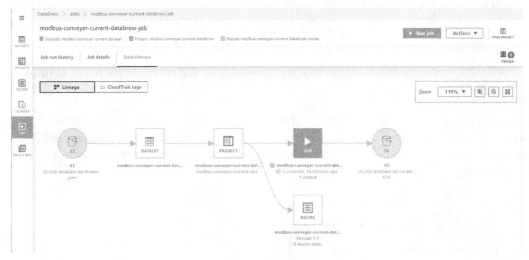

Figure 12.4 – The DataBrew process

The main goal is to move and transform data from the formatted data lake to the curated data lake. Let's explore the steps involved:

1. The formatted data lake data is a set of decoded individual JSON files containing a single measurement, which is not the optimum file storage format we would like. Thousands or millions of tiny data files cost more to move and store in S3, so this is something to consider when thinking about tiering your data in S3 over time.

2. `Dataset` is created, which collects the data in the formatted S3 bucket. We can determine how much data to collect and over what specific period. This could be useful later when we retrain the model with more recent data.

3. `Project` connects to the `Dataset` and allows us to visually work on the data to change it to the format we need. We can add a `Recipe` to the `Project`, which can subtly manipulate the data and transform it as required. We do not need any fundamental transformation, just some data aggregation into a single CSV file.

4. And finally, `Job` is created to take the resulting data and move it into the curated S3 bucket for use by the data science team.

While the functionality may be overkill for our purposes, it performs the task with minimal effort on our part. Along the way, we may want to explore the data within DataBrew to see whether we can assume a data engineer role and discover some new insights.

Figure 12.5 – The DataBrew project

The preceding image shows an example of the DataBrew project we have created, which provides a visual of the data stored in the data lake. To reiterate our data transform case, there are possibly dozens of ways to collect, view, and transform data. We are showing one possibility that is relatively easy to perform and repeatable.

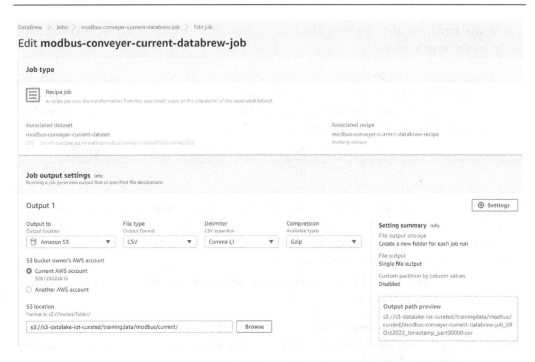

Figure 12.6 – A DataBrew job

Finally, let us look at the DataBrew job that stores the data in our curated data store. Using this general approach, we can provide the curated data to our data scientists or business analysts, who could use the data to perform analysis or ML. This can also help with security issues; by abstracting only the data that other teams will need to a different location, we can set stricter permissions on the formatted data lake as to who has access.

Note also that in *Figure 12.6*, we are prescribing how and where the data is stored. We define it as a single, comma-delimited CSV file stored within the training data directory in the curated data lake. Once the job is complete, data should be available and ready for use by the data team to train the model.

Setting up and training our model

We have been building and shaping our data model up to this point in the chapter, but now we have more decisions to make. Specifically, we must determine an approach when creating an ML model using SageMaker. Amazon SageMaker has several tools for working with ML and building models. When you launch SageMaker for the first time, it will insist that you set up a domain. This domain is where all of your persistent SageMaker resources are stored, including file storage, user and security information, and VPC configurations. Once this is complete, the list of ML environments will be available. *Figure 12.7* shows a view of the currently available ML environments. Depending on the practitioner's experience, skill, and needs, you may choose one of these options over the other to test ML scenarios.

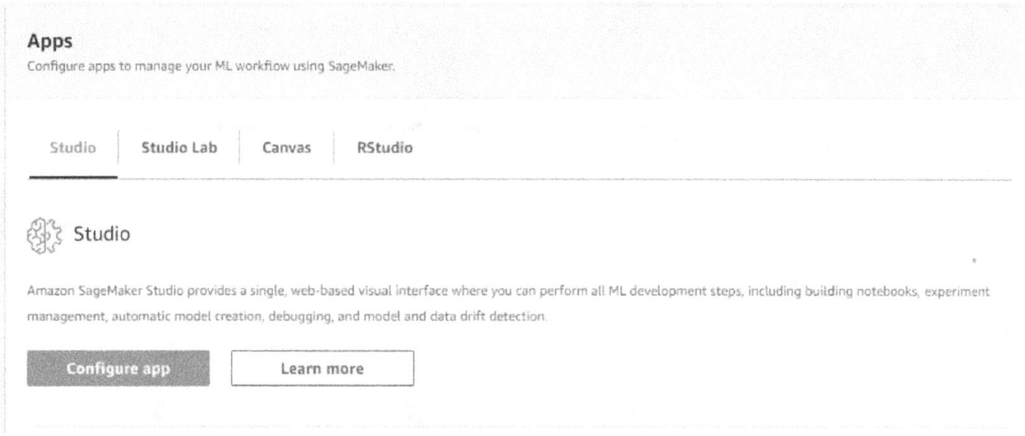

Figure 12.7 – SageMaker ML environments

Let's look at each option in slightly more detail:

- **Studio** is the option we will be choosing for our work. It is a fully featured workspace with an integrated Python SDK and the ability to build, test, deploy, and monitor your models. This may be the preferred option for developers or engineers who are used to programming in Python.

- **Studio Lab** is a free application that allows you to test your ML scenarios. It does not even require an AWS account. It may be a place to start if cost is an issue and you want to learn more about ML.

- **Canvas** is a no-code interface that allows you to use ML to generate predictions. Once models are created, you can import them into Amazon SageMaker for further collaboration and use.

- **RStudio** is an integrated development environment for R. For experienced or professional data scientists, this may be a preferred option. The statistical packages in R are generally more powerful than those in Python and are generally preferred by researchers and scientists.

Launching Amazon SageMaker Studio will send you into the studio workspace directly. From the launcher page shown in *Figure 12.8*, you can get a variety of samples or even available models. You can also launch SageMaker **JumpStart**, which has many examples, including the tutorial from which this example borrows heavily, called *An Introduction to SageMaker Random Cut Forests*. This tutorial provides a previously prepared dataset. In our example, we want to build the dataset from scratch using data collected from our local environment.

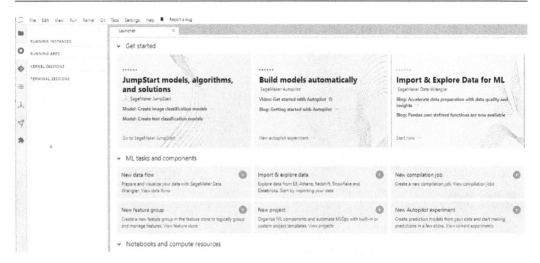

Figure 12.8 – SageMaker Studio

From the **File** menu, you can launch a new notebook. This is what is called a Jupyter notebook. Jupyter notebooks have been around for several years but have become popular with the advent of data science and data engineering goals. Unlike traditional IDEs, a Jupyter notebook allows you to combine documentation, code, and terminal output on a single page. This is powerful for learning and understanding your data and making comments or statements about how data should be transformed. Once you launch a new blank notebook, you can start to type in the provided transcript. Unlike other code samples, this one can be typed in and run in stages so that you can evaluate your efforts as you go.

Figure 12.9 – A Jupyter notebook

Figure 12.9 shows the starting point. To run a cell containing some code, highlight the cell and click on the run button (single arrow) at the top of the page. The output of the code will be displayed below the running cell. Let's review the code and understand the basics of building and deploying a model:

```
import boto3
import botocore
import sagemaker
import sys
from time import gmtime, strftime

# S3 bucket where training input and output will be stored.
bucket = sagemaker.Session().default_bucket()
prefix = "sagemaker/rcf-modbus"
execution_role = sagemaker.get_execution_role()
region = boto3.Session().region_name

# S3 bucket where the original testing data is stored.
downloaded_data_bucket = f"s3-datalake-iot-curated"
#downloaded_data_prefix = "testdata/current/modbus-conveyer-
current-databrew-job_24Sep2022_1664007637220"
downloaded_data_prefix = "trainingdata/modbus/current/modbus-
conveyer-current-databrew-job_09Oct2022_1665313856936"

# Output the data locations for reference.
print(f"Training input/output will be stored in: s3://{bucket}/
{prefix}")
print(f"Downloaded training data will be read from s3://
{downloaded_data_bucket}/{downloaded_data_prefix}")
```

Start with the usual imports you might see on any AWS code example. We want to define two S3 storage locations. The first element is where the algorithm will store the training data, and the second points to the data we created earlier in this chapter to train the model:

```
Training input/output will be stored in: s3://SageMaker-us-
west-2-305723022616/SageMaker/rcf-modbus
Downloaded training data will be read from s3://s3-datalake-
iot-curated/trainingdata/modbus/current/modbus-conveyer-
current-databrew-job_09Oct2022_1665313856936
```

Next, we will upload our data using the `pandas` data analysis library. Python programmers use this standard library to view and analyze data:

```
%%time
import pandas as pd
```

```
import urllib.request

data_filename = "modbus-conveyer-current-databrew-
job_09Oct2022_1664007637220_part00000.csv.gz"
s3 = boto3.client("s3")
s3.download_file(downloaded_data_bucket, f"{downloaded_data_
prefix}/{data_filename}", data_filename)
current_data = pd.read_csv(data_filename, delimiter=",")
```

Note the use of the magic %%time command at the start of this cell. This command prints the CPU time and the wall time for this cell. It provides an output of how much time and CPU the code in this cell took. Next, we download our data file and read our sample data into a pandas DataFrame. The default here is that there is no header on the data – that is, header=0 – but the read_csv function has many parameters to fit your file format:

```
CPU times: user 36.6 ms, sys: 0 ns, total: 36.6 ms
Wall time: 161 ms
```

We know what our data looks like from viewing it through AWS DataBrew earlier in this chapter, but we can perform several functions on the DataFrame to view the data in different ways:

```
current_data.head()
```

The DataFrame.head() function returns the first few rows of the data object. The default is five rows; however, this can be changed to fit your needs:

```
current_data.info
```

The DataFrame.info function provides a summary of the object, including the first five rows (head) and the last five rows (tail) of data. It also includes an overview of the number of rows and elements per row. *Figure 12.10* shows these two commands in action with the resulting views from each command.

```
: current_data.head()
```

	current	datetime	id
0	17	2022-09-23T07:22:04Z	824502
1	14	2022-09-23T11:12:15Z	824502
2	13	2022-09-23T09:56:12Z	824502
3	12	2022-09-23T13:20:22Z	824502
4	16	2022-09-24T04:35:06Z	824502

let's look at the data elements

```
: current_data.info
```

```
: <bound method DataFrame.info of         current         datetime    id
0          17  2022-09-23T07:22:04Z  824502
1          14  2022-09-23T11:12:15Z  824502
2          13  2022-09-23T09:56:12Z  824502
3          12  2022-09-23T13:20:22Z  824502
4          16  2022-09-24T04:35:06Z  824502
...        ...                   ...     ...
2646       14  2022-09-22T10:53:08Z  824502
2647       16  2022-09-22T11:22:09Z  824502
2648       14  2022-09-22T11:12:09Z  824502
2649       14  2022-09-22T11:42:10Z  824502
2650       17  2022-09-22T11:20:09Z  824502

[2651 rows x 3 columns]>
```

Figure 12.10 – Common DataFrame functions

It is also interesting to visualize the data if possible. `matplotlib` is a Python visualization library for creating graphs. This is probably an understatement, as Matplotlib is a powerful tool that can create publication-quality plots of your data:

```
%matplotlib inline
import matplotlib
import matplotlib.pyplot as plt
matplotlib.rcParams["figure.dpi"] = 100
current_data[500:700].plot("datetime", "current")
plt.ylim(0, 30)
```

The code for plotting a graph is pretty standard and easy to configure. We use the `pyplot` interface to generate a simple plot for our data. The last two lines of the cell are more interesting. Since our data is so large, we only look at the rows from 500 to 700. We can remove this constraint, but our data is not cleanly displayed without more configuration. The last row defines the range on the y index, providing some additional filters in the x-range to show a more straightforward overall evaluation. *Figure 12.11* shows the code sample and resulting plot of a subset of our data:

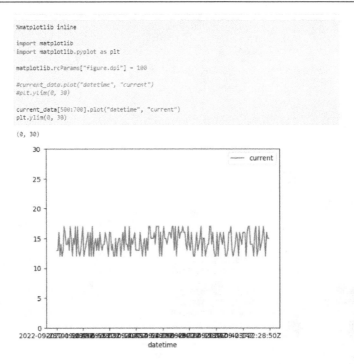

Figure 12.11 – A Pyplot of our data sample

This Pyplot sample is just a simple example to illustrate some of the things you can do with an interactive Jupyter notebook. It's a great way to have fun with your data and look at it differently. Next, we will set up our RCF by importing the library from SageMaker. We instantiate an instance with the necessary parameters to build our model:

```
from sagemaker import RandomCutForest
session = sagemaker.Session()

# specify general training job information
rcf = RandomCutForest(
    role=execution_role,
    instance_count=1,
    instance_type="ml.m4.xlarge",
    data_location=f"s3://{bucket}/{prefix}/",
    output_path=f"s3://{bucket}/{prefix}/output",
    num_samples_per_tree=512,
    num_trees=50,
    base_job_name = f"modbus-current-randomforest-
```

```
{strftime('%Y-%m-%d-%H-%M-%S', gmtime())}"
)


# automatically upload the training data to S3 and run the
training job
rcf.fit(rcf.record_set(current_data.current.to_numpy().
reshape(-1, 1)))
```

The parameters in the model development provide information on where and how to run the algorithm. As before, only two parameters can help shape the model – the **number of trees** and the **number of samples per tree**. There are many resources to share with you some ideas on calculating these values. At a minimum, we should ensure our dataset size equals num_trees * num_samples_per_tree:

```
Defaulting to the only supported framework/algorithm version:
1. Ignoring framework/algorithm version: 1.
2022-09-26 06:33:56 Starting - Starting the training job...
2022-09-26 06:34:24 Starting - Preparing the instances for
trainingProfilerReport-1664174036: InProgress
.........
2022-09-26 06:35:48 Downloading - Downloading input data...
2022-09-26 06:36:08 Training - Downloading the training
image..............
2022-09-26 06:38:50 Training - Training image download
completed. Training in progress..Docker entrypoint called with
argument(s): train
Running default environment configuration script


… lots of stuff going on here …


2022-09-26 06:39:29 Uploading - Uploading generated training
model
2022-09-26 06:39:29 Completed - Training job completed
Training seconds: 227
Billable seconds: 227
```

We have shortened the output you will see from this cell. It is a long sequence of commands and results, creating the model based on the training data. But at the end of the day, if you see the Training job completed output, you will know the job was completed successfully. Next, let us print the name of the training job for reference later:

```
print(f"Training job name: rcf.latest_training_job.job_name}")
```

This prints out the resulting job name that we generated:

```
Training job name: modbus-current-randomfore
st-2022-10-10--2022-10-10-07-13-57-548
```

Training jobs are stored within SageMaker. Run the command to print the latest training job to view the name. From within the SageMaker console, you can view the job and related information. Now that the training job is complete let's see how to test our model within the Jupyter notebook.

Deploying and testing the model

Remember, we used a reasonably homogeneous dataset to build our model. Our goal was to determine the mean and then use the model to check incoming data for future anomalies. Before we can test, we need to create an inference endpoint:

```
endpoint_name = f"modbus-current-randomforest-{strftime('%Y-%m-
%d-%H-%M', gmtime())}"
rcf_inference = rcf.deploy(initial_instance_count=1, instance_
type="ml.m4.xlarge", endpoint_name = endpoint_name)
```

The endpoint deploys the model to an instance where external users and code can access it. Instance size matters because this is a running instance where cost can be a factor. Less is more here if cost is a concern, and when is it not a concern? However, if your model is complex, performance could suffer greatly.

To interact with the endpoint, we need to define our data format by adding the ability for the endpoint to ingest a request and return results:

```
from sagemaker.serializers import CSVSerializer
from sagemaker.deserializers import JSONDeserializer

rcf_inference.serializer = CSVSerializer()
rcf_inference.deserializer = JSONDeserializer()
```

In this case, we define a serializer using a CSV format and a deserializer using JSON. Serialization allows you to extract meaningful data from an object or data structure, while deserialization does the opposite and creates a data structure of a specific type to return to the caller. Our input into the endpoint is a CSV file, producing a JSON object from the result.

Yeh! Our endpoint is ready to use. We can run a simple test:

```
import numpy
test_array = numpy.array([[8], [10], [12], [15], [17], [20],
```

```
[23], [25]])
print("Test Data: ", type(test_array), test_array)

results = rcf_inference.predict(
    test_array, initial_args={"ContentType": "text/csv",
"Accept": "application/json"}
)

import pprint
pp = pprint.PrettyPrinter(indent=4)
pp.pprint(results)
```

In this cell, we build a simple NumPy array. NumPy is a Python package that allows you to construct multidimensional array objects for scientific computing. In our case, it is only a single dimension with several values. We then print that test data to verify that it is correct.

Next, we use the `predict()` function to run an inference with our model. The model will take each value in turn and generate an anomaly score, as shown here:

```
Test Data:   <class 'numpy.ndarray'> [[ 8]
  [10]
  [12]
  [15]
  [17]
  [20]
  [23]
  [25]]
{   'scores': [   {'score': 3.9781670326},
                  {'score': 3.3770182451},
                  {'score': 1.0964158775},
                  {'score': 0.8990553522},
                  {'score': 1.1389528335},
                  {'score': 3.7042043919},
                  {'score': 4.3178147154},
                  {'score': 4.552415978}]}
```

Note that the higher the number, the more significant the difference from the mean of our test data. We built the model differently than we would in some cases. In some cases, we would use RCF to determine whether there were anomalies in our data. But in our case, we wanted to know what normal looked like and then use that to determine what was abnormal in the future. This could vary, of course, with most equipment. As measurements change over time, such as temperature or pressure, the model must be adjusted to accommodate these changes to look for spikes or drastic differences.

We could easily accommodate this same result with thresholds, but where would the fun be? Providing an easy-to-use single variable example reduces some challenges of learning ML as a first step. Next, let's see how to use this endpoint in real time.

Real-time inference

If you remember, earlier in this chapter, we added a placeholder in our decoding lambda code. That was specifically placed to allow us to leverage the model endpoint as real-time data is processed. The following set of code can now be added to reference our endpoint and evaluate incoming data:

```
### run against the model to calculate an anomaly score.
    data = {"current":str(event["current"])}
    payload = data['current']
    print("payload: ", type(payload), payload)
    runtime= boto3.client('runtime.SageMaker')
    response = runtime.invoke_endpoint(EndpointName="current-
rcf-2022-09-26-06-56", ContentType='text/csv',
Accept='application/json', Body=payload)
    print(response)
    result = json.loads(response['Body'].read().decode())
    print(result)


### Do Something here based on the resulting score?
```

We extract the current value in the preceding code snippet and create a simple payload. This payload is passed to the endpoint we created for the evaluation of new data. In the response of the API call, we will extract the body of the message and perform some action. Consider the results printed on the screen as shown here:

```
payload:  <class 'str'> 20
{'ResponseMetadata': {'RequestId': '87137003-3d4b-4656-86e3-
c00bd6ab8643', 'HTTPStatusCode': 200, 'HTTPHeaders': {'x-amzn-
requestid': '87137003-3d4b-4656-86e3-c00bd6ab8643', 'x-amzn-
invoked-production-variant': 'AllTraffic', 'date': 'Mon, 10
Oct 2022 16:31:47 GMT', 'content-type': 'application/json',
'content-length': '35'}, 'RetryAttempts': 0}, 'ContentType':
'application/json', 'InvokedProductionVariant': 'AllTraffic',
'Body': <botocore.response.StreamingBody object at
```

```
0x7fe96b6096a0>}
{'scores': [{'score': 3.7042043919}]}
```

From here, your imagination and requirements can run wild. Does a decision need to be made, an alert, or a system modification? You can see that other `lambda` functions can be created to leverage the endpoint API for any number of use cases.

We referenced the device's current because it led to an excellent and simple example that can be recreated relatively easily. The next step may be to build a model around three-phase voltage input to determine whether all the phases align and pull voltage within a given tolerance.

Before you finish, you should delete the endpoint. Especially if you are just running a test for learning purposes, run the following code in your Jupyter notebook to delete the endpoint:

```
sagemaker.Session().delete_endpoint(rcf_inference.endpoint)
```

This can either be done within the notebook or directly in the SageMaker console. Be sure to clean up all your resources and potential running instances. Amazon SageMaker runs instances under the covers for data wrangling, model building, and running endpoints. **The cost for complex data and models can be considerable if you are not prepared**.

Summary

ML is sometimes described as the ability of systems to learn and improve continuously. While, in a sense, it is true that models can be constantly updated and learned, we have not discussed that information here. Time and space in the book did not allow us to cover all the options, but we have provided some starting points to get you moving forward in a meaningful direction. One approach is to continually update your model and deploy it to your environment to ensure that as conditions change (think seasons), a model can be updated to accommodate these new conditions. As you become more comfortable with ML and using Amazon SageMaker, you can explore other opportunities to use ML successfully and continually improve it.

We have given you the basics and enough information to follow through the examples successfully, but not without a bit of work on your part. ML is a skill that takes time and effort to perform successfully, but it's not impossible. As you can see from this chapter, you can start with a basic model and grow as your skills and experience build.

As this is the final chapter, we hope you have enjoyed and learned from this book as much as we have writing it. There are so many paths you can take using different AWS services or third-party solutions that it is hard to know the best approach. One piece of advice is consistency. Depending on consistent services, procedures, and standards is always helpful. Try different techniques, consider future needs, choose a direction, and go!

We wish you good luck in your IIoT journey!

Appendix 1

General Cybersecurity Topics

While we have discussed the topic of *secure by design* in the context of industrial control systems, we would like to provide additional thoughts around IT and **Operational Technology** (**OT**) cybersecurity from *Chapter 5, OT and Industrial Control Systems*, and *Chapter 6, Enabling Industrial IoT*.

The importance of security in manufacturing facilities cannot be over-emphasized as the world braces against cyber threats and attacks. As we have covered, IT incidents result in application outages. In contrast, OT incidents result in real-world consequences, potentially life-threatening accidents, or production and revenue losses.

The challenge here is that we are delving deeper into OT and connecting to systems within this book. This effort comes with risk and the need to mitigate that risk as much as possible with the intrusion of IT into the OT domain. As an architect, the last thing you want to do is introduce hazards or new security holes when adding new capabilities.

OT security strategies

Networks within OT environments can be some of the most insecure components within an organization. Cabling runs from one control box to another control box, providing a backbone of communication for machines and operations. However, sometimes, little thought is given to securing that network or ensuring that communication is managed effectively. This is not always the case, and if your organization has embarked on locking down OT networks, then you are ahead in the game.

Nonetheless, legacy environments often just run – a key metric is keeping them running and producing results as efficiently or quickly as possible. Little thought is sometimes given to the potential threat lurking within the cables connecting our equipment.

IEC 62443 is the international standard that delves into cybersecurity for OT and industrial communication networks. This is an excellent benchmark to consider when understanding policies, procedures, system requirements, and components for building a defense-in-depth strategy for the enterprise. NIST also provides SP 800-82 standards for **Industrial Control Systems** (**ICS**) security. Both of these industry-leading standards can be referenced to help you create a secure system design.

There are multiple strategies to implement security in the OT realm – and there is no one best way, as there are numerous factors involved, the age of the plant, brownfield/ greenfield, the immediate project budget, and so on. These two prominent industry-leading standards consider comprehensive and robust steps as a process book to build a future-ready and attack-resilient security mechanism. They are as follows:

1. **Defense-in-depth** integrates people, technology, and operations to establish variable barriers to prevent single points of failure. It also assumes that the source of threat may not be the only one that is an essential factor to consider during system design. This strategy creates multiple protection sheaths to prevent attackers from launching attacks, but once inside the perimeter, implicit trust is assumed, creating a possibility for attack exploitation. Hence, we also have the evolutionary strategy of zero trust.

2. **Zero-Trust Security Architecture (ZTA)** has also been growing in relevance in the OT domain, given its popularity in the IT world. This security paradigm focuses on resource protection based on the premise that authorization decisions are made closer to the requested resource and are continuously evaluated rather than implicitly granted. This strategy paves the way for resources to authenticate every time, regardless of importance, thus preventing default access patterns and horizontal network traversals. Still, the applicability of ZTA to legacy PLC devices and systems can be challenging for two reasons. First, they may not support a full-blown implementation of ZTA; the other is access latency within the network, which can be a dampener. Thus, a specific combination of defense-in-depth and zero trust must be employed to arrive at a suitable OT security architecture for a given scenario.

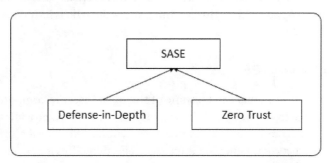

Figure A.1 – SASE = Defense-in-depth + zero trust

Figure A.1 illustrates the relationships within this standard. This is wrapped up within a **Secure Access Service Edge (SASE)** (pronounced sassy) framework. Gartner defines SASE as a robust framework and set of capabilities for providing access and remote connectivity to a network.

Comprehensive security framework

Security is a continuous process and needs to be engrained by design into people, processes, and plants using tools and technology to create a piece of well-oiled machinery against the growing cyber-attack landscape. *Figure A.2* depicts the five cyclic and perennial processes within the cybersecurity framework:

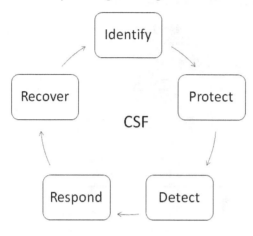

Figure A.2 – Cybersecurity framework

The five processes, **Identify**, **Protect**, **Detect**, **Respond**, and **Recover**, apply to IT and OT domains. However, selective and adaptive applications are deemed necessary due to the fundamental difference in the design, equipment, and priorities.

As depicted in *Figure A.3*, we can become familiar with the layers of the ISA-95 architecture and the distinction between OT and IT. Within each layer, there are different networks. Starting from the bottom, layer 1, the production network interlocks all the machines. Then, we have the control network at level 2 with PLCs, RTUs, terminal units, etc. On top of this is the automation network consisting of SCADA, HMI, the historian, the IoT platform, and so on.

Then, we have the critical **Industrial Demilitarized Zone (iDMZ)**. This essential layer, also called a landing zone, completely isolates the OT and IT network. The iDMZ is flanked by firewalls on either side, and they restrict and direct data movement between the two domains.

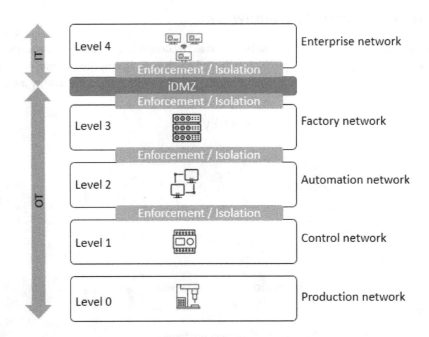

Figure A.3 – Security across various networks in a plant

A well-structured **Demilitarized Zone (DMZ)** is of utmost importance as we speak of a secure layer that shields the production network from any attack on the IT network. A simple phishing email click-baited by a remote contractor in one corner of the world can provide a backdoor for unlimited access to any production PLC across any of the plants to a hacker from a virtually undetectable location.

This sounds more than scary. A well-designed DMZ creates the much-needed isolation between these networks to prevent or make this kind of scenario difficult. There must be sufficient enforcement isolation through firewalls and logical segmentation implemented across all the layers. This is an upfront investment considering the overall security of the enterprise.

Orchestrating a robust security program

Having realized the importance and complexity of an OT security program, there must be a structured process for implementing the program across the enterprise. It starts with change management and new security organization creation headed by a **Chief Information Security Office (CISO)**. They then head the **Network Operations Center (NOC)** and **Security Operations Center (SOC)**. These teams work towards the singular goal of ensuring that business operations are sustained smoothly. Individually, these teams are different. While the SOC is primed for securing and protecting the organization against cyber threats, the NOC is responsible for ensuring that the network infrastructure is consistently maintained to sustain business operations.

Without delving deeper into the topic, we would like to stick with the singular objective of presenting to the readers the options and strategies available to get started on this path. OT cybersecurity can be divided comprehensively into five distinct pillars. Each of them uniquely adds layers of security to the system. These pillars are represented in *Figure A.4*:

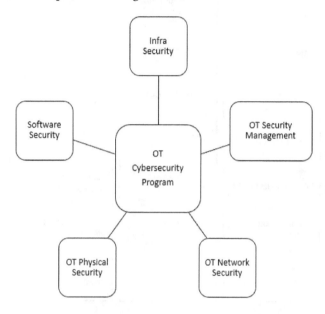

Figure A.4 – Security across various networks in a plant

The first pillar, **Software Security**, uses tools to harden and protect devices and end-user applications. It also has defined process flows for anti-virus signature updates, security patches, OS updates, and more. **Infra Security** protects servers, switches, and factory devices in the OT network. **OT Security Management** defines, in general, the functions of the SOC and the tools that aid in security conformance and maintenance. **OT Network Security** is mainly the NOC functionalities used to monitor the network, detect unusual traffic and accesses, and manage these. Last, we have the OT Physical **Security** pillar concerning the surveillance of locations in the server room, sensitive computing machinery, etc.

There are many areas to consider when considering IT/OT security and potential attack surfaces. We have outlined some areas to consider as you define changes to your network architecture in *Figure A.5*. These guide each of the pillars and the nature of applications and solutions that are needed.

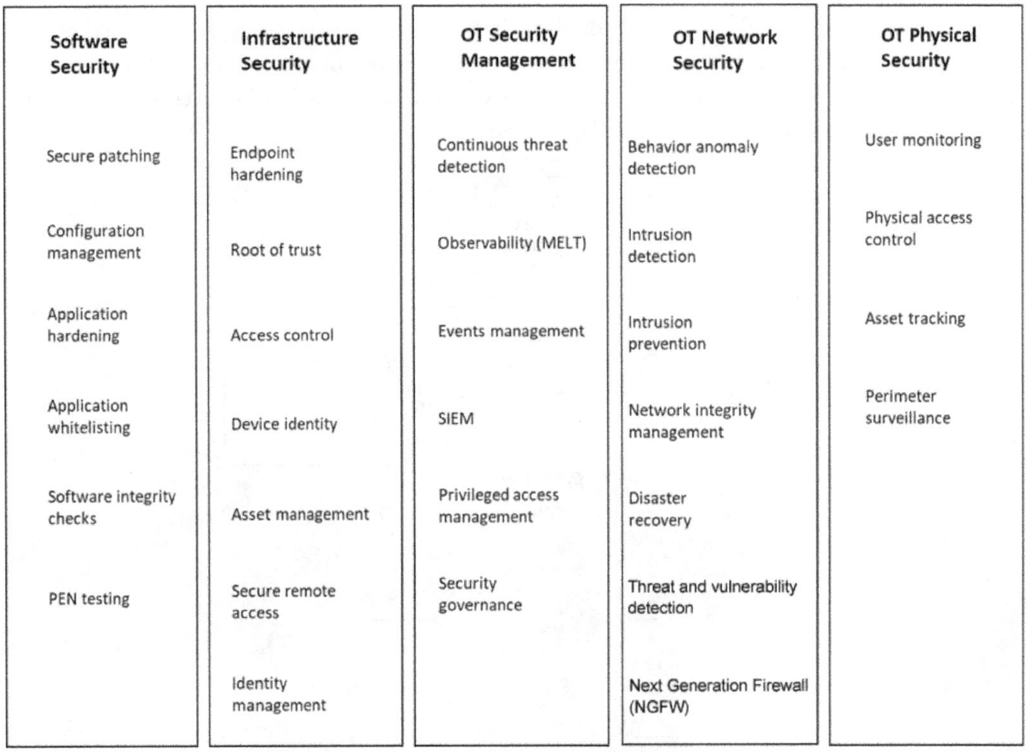

Software Security	Infrastructure Security	OT Security Management	OT Network Security	OT Physical Security
Secure patching	Endpoint hardening	Continuous threat detection	Behavior anomaly detection	User monitoring
Configuration management	Root of trust	Observability (MELT)	Intrusion detection	Physical access control
Application hardening	Access control	Events management	Intrusion prevention	Asset tracking
Application whitelisting	Device identity	SIEM	Network integrity management	Perimeter surveillance
Software integrity checks	Asset management	Privileged access management	Disaster recovery	
PEN testing	Secure remote access	Security governance	Threat and vulnerability detection	
	Identity management		Next Generation Firewall (NGFW)	

Figure A.5 – Security across various networks in a plant

The ecosystem, though seemingly complex, comes together as an orchestrated engine, providing the much-needed base for creating an architecture that is secure, robust, and scalable for building world-class IIoT solutions.

OT cybersecurity areas of concern

In addition to the preceding list, *Figure A.6* provides a short list of some of the pillars of cybersecurity by type. These can be considered capabilities of a fully functional DMZ, as discussed earlier. Factory personnel use applications deployed within the OT network or factory network almost exclusively for viewing, gathering or controlling data. The ability to update and access those applications securely is critical.

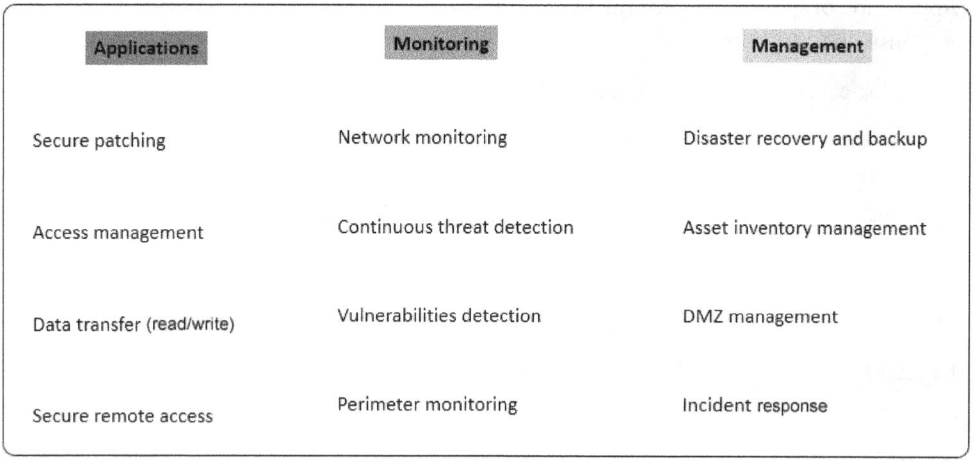

Applications	Monitoring	Management
Secure patching	Network monitoring	Disaster recovery and backup
Access management	Continuous threat detection	Asset inventory management
Data transfer (read/write)	Vulnerabilities detection	DMZ management
Secure remote access	Perimeter monitoring	Incident response

Figure A.6 – Major pillars of OT cybersecurity

Network and equipment monitoring is another area where capabilities can be added. The ability to continuously monitor threats to the system can provide some initial protection you may not already have.

Finally, the management of the network and environment should be considered. How are firewall rules checked and changed? What processes are in place for incident response? Is there a **Disaster Recovery (DR)** plan in place?

Figure A.6 can act as a starting point. Ask any or all of these questions to OT engineers or IT networking architects. How do we perform perimeter monitoring? This gets everyone thinking in the same way and working together to put a more robust set of tools and processes in place.

OT cybersecurity best practices

We want to list some of the industry-leading best practices regarding OT cybersecurity. These are by no means exhaustive, and the nature of adoption depends on each solution and its implementation:

- The strong presence of a continuous security program
- Commitment from all stakeholders to uphold the best practices for security
- Knowledge sharing and security training for all stakeholders
- Standard network design with structure documentation
- Clear inventory of assets and a program to refresh them periodically (hardware and software)
- High availability and resilience by design
- The DMZ is the only mechanism for data traversal from IT to OT networks and vice versa

- Application of specific tools for the five CSF processes of identification, protection, detection, response, and recovery

- Use of logical network segmentation and isolation where possible

- Implementation of multi-factor authentication, zero-trust, and defense-in-depth strategies for user access and remote access

- Clear disaster recovery and a business continuity plan in place

- No inbound connections directly to automation networks

- Monitored outbound with privileged access management

- Regular system backup and restoration mechanisms

Summary

We have only scratched the surface of these topics. Still, at a minimum, we wanted to provide an overview for thinking about and designing secure operations at a manufacturing facility. Some lead-in skills may be necessary to take this further and better understand the information. Security and networking fundamentals are required when considering this topic.

It is also required that IT networking and OT engineering work closely to understand the impact that changes to the network will have. Wrapping the entire network in a big firewall is an excellent first step. Then a structured approach to applying security at various layers can be considered.

Hopefully, this short appendix gets you thinking about approaches you may consider and, more importantly, moving in the right direction.

Index

`Packt.com`

Subscribe to our online digital library for full access to over 7,000 books and videos, as well as industry leading tools to help you plan your personal development and advance your career. For more information, please visit our website.

Why subscribe?

- Spend less time learning and more time coding with practical eBooks and Videos from over 4,000 industry professionals

- Improve your learning with Skill Plans built especially for you

- Get a free eBook or video every month

- Fully searchable for easy access to vital information

- Copy and paste, print, and bookmark content

Did you know that Packt offers eBook versions of every book published, with PDF and ePub files available? You can upgrade to the eBook version at `packt.com` and as a print book customer, you are entitled to a discount on the eBook copy. Get in touch with us at `customercare@packtpub.com` for more details.

At `www.packt.com`, you can also read a collection of free technical articles, sign up for a range of free newsletters, and receive exclusive discounts and offers on Packt books and eBooks.

Other Books You May Enjoy

If you enjoyed this book, you may be interested in these other books by Packt:

Industrial Digital Transformation

Shyam Varan Nath, Ann Dunkin, Mahesh Chowdhary, Nital Patel

ISBN: 9781800207677

- Get up to speed with digital transformation and its important aspects
- Explore the skills that are needed to execute the transformation
- Focus on the concepts of Digital Thread and Digital Twin
- Understand how to leverage the ecosystem for successful transformation
- Get to grips with various case studies spanning industries in both private and public sectors
- Discover how to execute transformation at a global scale
- Find out how AI delivers value in the transformation journey

Building Industrial Digital Twins

Shyam Varan Nath, Pieter van Schalkwyk

ISBN: 9781839219078

- Identify key criteria for the applicability of digital twins in your organization
- Explore the RACI matrix and rapid experimentation for choosing the right tech stack for your digital twin system
- Evaluate public cloud, industrial IoT, and enterprise platforms to set up your prototype
- Develop a digital twin prototype and validate it using a unit test, integration test, and functional test
- Perform an RoI analysis of your digital twin to determine its economic viability for the business
- Discover techniques to improve your digital twin for future enhancements

Packt is searching for authors like you

If you're interested in becoming an author for Packt, please visit `authors.packtpub.com` and apply today. We have worked with thousands of developers and tech professionals, just like you, to help them share their insight with the global tech community. You can make a general application, apply for a specific hot topic that we are recruiting an author for, or submit your own idea.

Share Your Thoughts

Now you've finished *Industrial IoT for Architects and Engineers*, we'd love to hear your thoughts! Scan the QR code below to go straight to the Amazon review page for this book and share your feedback or leave a review on the site that you purchased it from.

`https://packt.link/r/180324089X`

Your review is important to us and the tech community and will help us make sure we're delivering excellent quality content.

Download a free PDF copy of this book

Thanks for purchasing this book!

Do you like to read on the go but are unable to carry your print books everywhere? Is your eBook purchase not compatible with the device of your choice?

Don't worry, now with every Packt book you get a DRM-free PDF version of that book at no cost.

Read anywhere, any place, on any device. Search, copy, and paste code from your favorite technical books directly into your application.

The perks don't stop there, you can get exclusive access to discounts, newsletters, and great free content in your inbox daily

Follow these simple steps to get the benefits:

1. Scan the QR code or visit the link below

https://packt.link/free-ebook/9781803240893

2. Submit your proof of purchase
3. That's it! We'll send your free PDF and other benefits to your email directly

www.ingramcontent.com/pod-product-compliance
Lightning Source LLC
Chambersburg PA
CBHW062057050326
40690CB00016B/3124